The author and his wife.

A ZOO WITHOUT BARS

A Zoo without Bars

Life in the East African Bush 1927–1932

by

T. A. M. Nash, C.M.G., O.B.E., D.Sc.

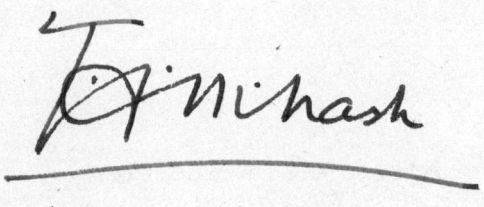

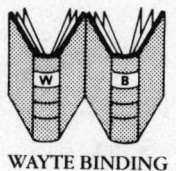

WAYTE BINDING

FOR WENDY

(The Smallest Empire Builder)

ISBN 0 9509730 0 9

Published by and obtainable from
Wayte Binding
97 St James Park
Tunbridge Wells
Kent TN1 2LQ

Produced for the publisher by
Chambers Green Ltd.
Tunbridge Wells, Kent

Printed in Great Britain

CONTENTS

Preface

An attempt has been made to record day-to-day life in bush during the Colonial Period. I had 32 years of African service, 1927–1959, of which the first 22 years were spent in the field, studying tsetse flies in relation to the diseases which they carry and their control; the scientific side of this work has already been published. Here, the intention is to describe the *background* to the first five years of my service in Tanganyika Territory, now Tanzania.

This book is intended for the reader who is interested in the living conditions, the wildlife, the peasant and the European characters who gave him so much to laugh at. What I have written is well documented being based on 117 letters home and 285 pages of descriptive accounts. Throughout, I have refused to pander to the modern sensibilities of the politically-minded, be they black or white; in the years covered here, we thought that we were building the second course of the Empire and never dreamt of independence in our time.

The incidents recorded have been related in chronological order so as to provide variety in subject matter, and also to portray the evolutionary changes in the author's youth from loneliness to company, from living in a mud hut to a mud house, from inaccessibility to accessibility, from bachelorhood to marriage and a baby – all in an area teeming with game.

The value of this record is that it is buttressed with documents and not by contorted memories. Having recently read the letters for the first time since they were written over 50 years ago, I was intrigued by long-forgotten incidents and shocked by the inaccuracies of those that 'I clearly remembered'. Any errors in this book are regretted, but they will stem from my ignorance at the time of writing home. My beginner's inefficiency as a hunter of dangerous game will evoke mirth among the experts, however it is hoped that first impressions and the reality of bush life in those days will be of general interest.

No attempt has been made to change the names of the characters mentioned, except in a few instances where my comments would give offence. I feel that the real names may be of minor historical interest.

Acknowledgements

I am most grateful to Sir Rex Niven for helpful suggestions concerning the manuscript and to Dr. A. M. Jordan, my successor as Director of the Tsetse Research Laboratory at Langford, for his aid on various matters. Dr. and Mrs. P. A. Langley also provided valuable comments on the manuscript. I am indebted to Noel Vicars-Harris, who has also refreshed my memory on various details; but, as will be apparent to the reader of this book, my gratitude to him and Maria extends to my earliest days in Africa. Lastly I am indebted to Gill Wilkinson and Mary Studley for secretarial assistance, to Malcolm Parsons who performed wonders with my ancient negatives and prints, and above all to John Wayte who arranged for the publication of this book.

Map of Africa circa 1928

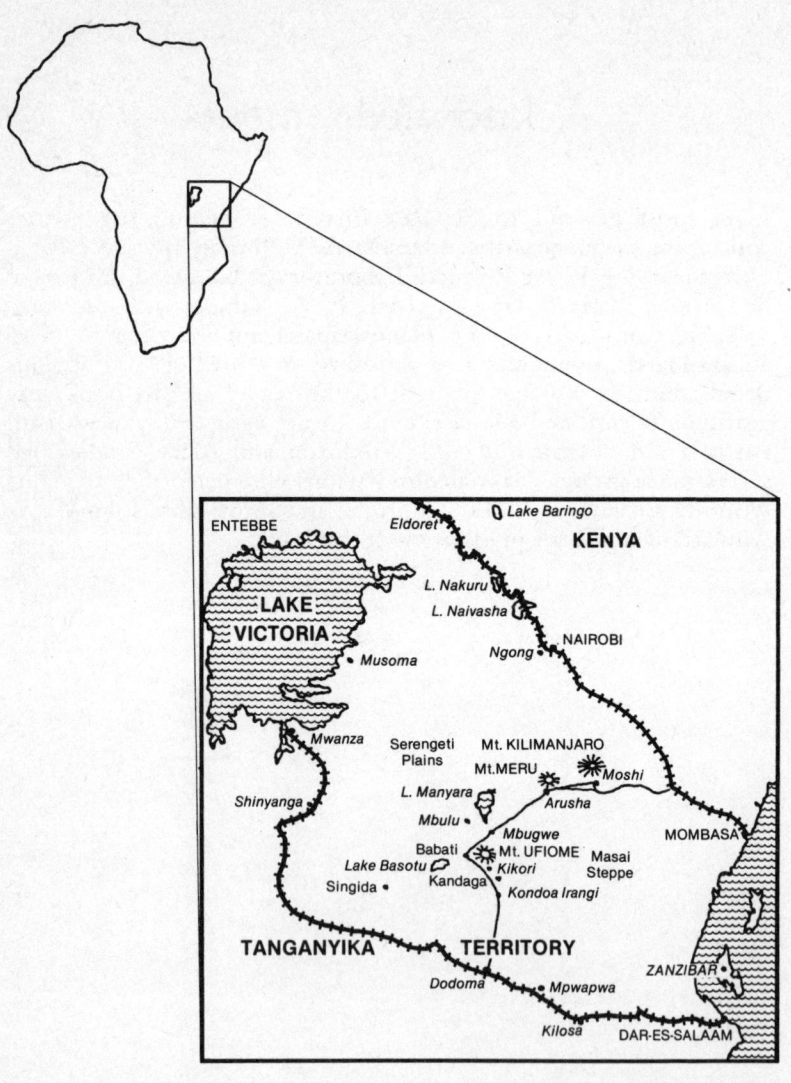

FIRST TOUR

CHAPTER 1

Appointment and Journey to Kondoa Irangi

I was dissecting an insect when Spinks, the laboratory assistant appeared. 'What 'ave you been up to, Sir? Prof. wants you pronto.' I knocked on the door and went in. Professor Munro waved me to a chair. 'This is Mr. Swynnerton, head of the Game Preservation Department and Tsetse Fly Control in Tanganyika Territory. He wants an entomologist.' I took a good look at the visitor: tall and wiry, wavy red hair going greyish: twinkling green–blue eyes that put one at ease: reddish hair and freckles on his hands. Tough though he was, he was obviously sensitive and lit by a flame.

My family had long been imbued with a sense of service to the Empire. One of my grandfathers had been a Surgeon Major in the West Indies; the other, a chaplain in the East India Company had, with my grandmother, been through the Mutiny. My father, a Colonel in the Royal Army Medical Corps, had been many years in India where I spent my early childhood. But I was tired of India: the drawing room was cluttered up with Benares brass, clay gods and curios. I was intrigued by Darkest Africa with its great herds of game. Recently I had read John Masefields's 'Multitude and Solitude', a slight novel in retrospect, but with graphic descriptions of sleeping sickness in man, and 'nagana' in cattle; both diseases are carried by tsetse flies. Why not go out and work on the problem? I accepted Swynnerton's offer, applied to the Colonial Office for the post, and fortunately got it.

In September 1927, when just 22, I sailed from Tilbury on S.S. Malda, with my mother and Wendy, my fiancée, waving from the quayside. It was a British India Line ship, a Line aptly referred to as the Bug and Insect, as exemplified by the then current seaman and Purser story.

'I want to complain, Sir, about the bugs on this ship.'

'I can assure you there isn't a single bug on this ship.'

'I quite agree, Sir, they're all married with children.'

Our first stop was at Marseilles where we spent the night unloading. After breakfast the passengers were leaning over the rails waiting for the gangway to be raised, when a little French girl ran up it crying, 'Mr. Coop-père, Mr. Coop-père. Where is Mr. Coop-père?' Eventually she found him cowering in a corner and said, 'Here is your passport and wallet which you left in my room last night.' Thereafter his name ceased to be Cooper.

At Port Said I bought some Turkish Delight which I put in a drawer in my cabin; next morning it was swarming with red ants. I rang for the steward. 'Yes Sir, we will soon fix them.' He returned with a jar of chopped raw meat and a toilet roll. 'Now Sir, tear off a piece of paper, put some meat on it, shut the drawer. Inspect whenever you enter the cabin. Roll up the paper, meat and all, and throw it out of the porthole.' It worked. I had had my first practical lesson in insect control. (No slur is implied on the B. I. Line. Before the introduction of modern insecticides such incidents were inevitable.)

Port Sudan lived up to its reputation of being the hottest place in the Red Sea. After sunset the cabin temperature was 108°F. Next morning when I went for my bath, the water was so hot I could barely put my foot in it. I drained the bath and turned on the cold tap, but it was just as hot.

The ship came to life again after entering the Indian Ocean. Dallas, who had lost a leg in the war, joined in a game of rugger on the deck, and a doctor broke his arm. We were all wearing dinner jackets. It was a riotous evening.

After a 5 week voyage we reached Dar es Salaam.

A letter from Swynnerton awaited me. I was to break my train journey at Kilosa, report to Captain Fairweather, Senior Game Ranger, pack up scientific equipment, and catch the next train for Dodoma whence a lorry would take me up to Kondoa Irangi. As there was no train until the Monday, I had to spend three nights in the old German hotel – renamed The New Africa. The only incident was seeing a lady off. She got into her car, backed it into the snake charmer's basket, cobras shot out, I shot into the hotel, and she unwittingly drove off.

I de-trained at Kilosa – a single platformed station. There was no one to meet me. The Indian Station Master pointed out a white house on a hill. I walked up the steeply winding road in my white Bombay bowler (topee) and tusser-silk suit, until I reached the house at about 10 a.m. Two huge bleached elephant skulls, mounted on plinths, guarded the wide steps up to the front door. I climbed them. A voice said 'karibu' – come in. I entered. There were two men in dinner jackets, Captain Fairweather and Captain Micky King, a tsetse reclamation officer who had been in bush for eighteen months. They were having breakfast after a wild night at the District Officer's house. After breakfast they changed, and then took me down to the Game Office. I soon packed up the equipment and we went to a Greek bar for drinks. Later, somewhat the worse for wear, Fairweather drove us back for lunch. The hill was steep. He missed the corner and went over the side. We climbed out. Fairweather said, 'That's all right, it often happens. The convicts will pull it out during lunch.' They did.

After lunch, Fairweather said, 'I know what we'll do with the boy. We'll take him along to the Baron's place.' We arrived in a native village. There, sitting under a huge fig tree, was a white man with a towel round his waist. He rose. I was introduced. He shook hands in the most courteous fashion and said, 'Excuse my attire, but I have a dose of clap.' After a few drinks, a large black mammy waddled up. The Baron arose, bowed and said, 'Allow me to introduce Mr. Nash – the Baroness.'

Micky King had a very bloodshot eye which he kept rubbing with a dirty hankerchief, a spark from a bush fire had got into it. He was also going to Kondoa, and catching the Dodoma train which was due very early the next morning. Fairweather had arranged for the driver to sound his whistle on approaching Kilosa. At 1.30 a.m. I awoke to hear the whistle screeching in the valley. Fairweather, in pyjamas, drove us down the winding zig-zag road at a mad rate, sounding his claxon horn all the way to warn the station master to keep the train, but as the Station Master said, 'When zee train is 16 hours late, what matter, another half.' We leapt into a compartment and the train pulled out. Micky was in a bad way, sleeping-drunk, constantly groaning and rubbing his eye. Twice during the night I washed it out with boracic acid lotion.

On reaching Dodoma, his servant helped him across to Georgiou's hotel, whilst I collected our luggage. On rejoining Micky, he was saying to Georgiou, 'If you don't bloody well find me a room with two beds in it I will break every bone in your ----- body.' We had a grand room in two minutes. I got very fond of Micky, but felt like the ship's cabin boy with the choleric captain. 'Nash! your bill is going on mine; you don't have to pay a cent here.'

I expostulated. 'What the hell do you mean boy. Obey orders.' (Two years later I was at Georgiou's hotel. It was crowded with people, sitting outside around tables, drinking, whilst awaiting the train; a tame greater kudu calf was going round eating the cigarette ends. I suddenly saw Micky, wearing a black eye shade, and very drunk. He spotted me, staggered to his feet and shouted, 'Silence. This boy saved my life. How shall I repay him? I know - I'll give him a free hair cut. Barber! Barber! Later, I heard that he had gone in for gold mining, had fallen down his mine shaft and broken his neck.)

An Albion lorry had been sent to meet us, but next morning I was the sole passenger, as when the doctor saw Micky's eye he packed him off to Dar es Salaam hospital. What a joy it was to return to reality, sitting for 7 hours beside the African driver for 100 miles of bone-shattering earth road. The back of the lorry was laden with baggage, natives and their wives. We stopped on every hill-top to pour water into the radiator.

We filled our water cans at Kirema Drift – a wide, sandy river crossed by a concrete drift. (On a later occasion I found a raging torrent. There was one tent on the camping site, occupied by Crowther, a road superintendent who was very drunk; after dinner he was worse. It was an awful night. The lions roared incessantly; they were hungry as the game was on the other bank. The boys kept throwing logs on the fire. I slept fitfully with a rifle inside my mosquito net, and left the hurricane lamp alight. My dog's ever moving head indicated where the lions were. Next morning when a bleary-eyed Crowther appeared, we had a shaggy-dog type of conversation.

'What a fiendish noise those lions kicked up.'

'There ain't no lions here.'

Come and look.' I pointed out pug marks between the guy ropes of his tent.

'That's not lion spoor.'

'Then what the hell is it?'

'Why, it's a lion-ess.')

To return to the journey, we left the Kirema, climbed the escarpment, and drove through stony hills and parkland, dotted with euphorbia and baobab trees, to reach Kondoa at about 4 p.m. I was yearning for a cup of tea, but never got one.

My first impression of Tanganyika Territory was that the Europeans were either mad or drunk, but wider experience indicated that I had had an unusual introduction.

CHAPTER 2

Apprenticeship

The lorry drew up at a house which was being used by Swynnerton as a temporary headquarters and as a dormitory for his young men coming in from bush to report. It was a Friday.

Having greeted me, Swynnerton immediately started talking shop. 'Here is a drum of castor oil and a keg of resin. I want you to prepare a birdlime that will hold tsetse when they settle on it. There are the flies. Tomorrow I will give you some ox intestines, and some blood that has been defibrinated so that it won't clot. I want you to add formaldehyde at a dosage which will kill the flies but not deter them from probing. We leave at 1 p.m. on Sunday. Before departure you will fill a length of intestine with the blood mixture and tie it zig-zagging onto the bars of this frame. On the other two, nail squares of antelope hide and paint them with birdlime. Then get the driver to lash all three frames onto the tail-board of the lorry.' I started work at once.

On Sunday, having postponed his departure by a day, Swynnerton took me to a sundowner (drinks party) to meet the other six European inhabitants of Kondoa. It was very pleasant, with drinks in moderation.

On Monday the great day came. The lorry was so packed with loads that I was allowed only one suitcase and a valise, but no servant. Swynnerton sat next to the driver, armed with pencil and notebook. I sat at the tail-board watching the screens and ready to shout out, 'One fly feeding on guts, Sir; three stuck to birdlime.' But the great moment never came. Soon we hit an imperial pot-hole. The guts broke. I was covered in blood and formaldehyde. I persevered. Ten minutes later we hit the father and mother of all pot-holes; the bottom lashing broke, the screen flew up and hit me in the face, to which birdlime was now added. But Swynnerton was an absolute gentleman: he let me sit beside him for the rest of the

17

journey. We reached Singida at 11 p.m., having had nothing to eat since lunch, apart from 'Nice' biscuits – thin, sugar-coated, coconut flavoured.

Singida lies in a plain, studded with gigantic boulders piled up on each other to great heights, looking from a distance like ruined castles. There are two huge lakes. One of them held two flocks of flamingoes, beautiful birds with their white backs and pink breasts, as seen when standing fishing. Suddenly, they all taxi along the surface, gaining speed by beating the water with their feet; then gradually they rise and sweep through the air, flying with long necks strained forwards and stiff legs stretched out behind. They turn so many colours in the evening sun. But it was not on this occasion that I watched them, as we left at dawn for Matelele.

The dry season roads in this area followed old elephant paths and wound in and out. They were so narrow that thorny branches brushed each side of the lorry. The road surface was ghastly; every few yards the driver would jam on the brakes, sidle over a bump, accelerate, then brake – the average speed was 10 m.p.h.

On the way to Matelele we passed within 20 yards of a herd of giraffe. Rigid and motionless, they are so well camouflaged that you do not see them. Alarmed, their necks shoot up; they are rigid. Then their nerves give way and they bolt. The herd of Noah's Ark horses on stilts thunders away yet somehow they never fall over their own feet. We reached Burtt, the botanist's camp, beautifully situated on a cliff overlooking the dried up bed of a river. Vegetable Ivory palms lined the sandy banks, with straight trunks swelling into bulges below the crowns; the crowns were at eye-level from our tents.

On this 9 day *safari* with Swynnerton we lived on tinned food, as he did not want to waste room on the lorry for a cook and his utensils. For two nights I shared a hut with him. I fell asleep to the hammering of his typewriter. Several hours before dawn I was awoken by something falling on the ground. 'I haven't woken you, have I Nash?' 'It's all right, Sir.' He then talked shop until dawn. His boy brought two cups of tea and a tin of 'Nice' biscuits. We were to have breakfast on the road; after walking ten miles we got it – 'Nice' biscuits. After a further fifteen miles we got back to camp and I opened a tin. That night it was the same story, but I did not admit that I had heard something fall.

I later learnt that Swynnerton never slept for more than four hours a night and was always in too great a hurry to stop for food. He was a delightful person, a tremendous enthusiast, but utterly exhausting. His native name was '*bwana, funga-funguwa*' (Master, pack – unpack).

After a night at base, I and Duncan, a reclamation officer who had come out on the previous boat, were sent out to camp on the far bank of the Bubu river, some six miles west of Kondoa. I was to continue the work on 'blood and guts'. We just had time to get our tents up before dark but not time to collect firewood, which was a pity as the place was stiff with lion. My *askaris* and servants slept under the tent flaps so as to be near the two hurricane lamps which I kept on all night. What with the grunts and gurgling noises they made in their sleep and the wildlife noises outside, it was a boisterous night.

I had taken on a servant, Juma by name, in Dar es Salaam. He was not a great success but meant well. Every time I gave an order he would say 'Yes! *bwana*', 'Yes! *bwana* and dash off before I had finished speaking, only to do the wrong thing. On one occasion he filled the bottom of my filter as well as the top with dirty water. My cook was old and scruffy, with a villainous face, but he was a good cook. Local food was very cheap with chickens at 6d each and eggs at 50 for 8d, but both chickens and eggs were very small. Tinned food bought at Kanji Damani's shop in Kondoa was beyond my means at 5 shillings for a 2 lb. tin of biscuits or for a tin holding 5 sausages. In any case, unless one could eat 5 sausages at a sitting, the left-overs would soon have gone bad with the temperature rising to 115°F in the tent.

Not far away, reclamation work was going on with labourers cutting down woodland to prevent an advance by tsetse; Noel Vicars-Harris was in charge of the scheme. He had been supervising a cotton plantation in Brazil, where he had recently married Maria, a charming 17 year old girl. We became great friends, and often went hunting for meat for his labour gang. On one occasion we went down to the Bubu river. We separated. I shot a bush fowl, a partridge-like bird, but saw no antelope. Noel shot a waterbuck, and shortly afterwards saw what he thought was a female reedbuck on a bush-covered hillock some 15 yards away - until he saw a waving tail, then a lioness, cub and black maned lion. He only had a .256 rifle so wisely did not fire, but made Maria retreat. The lioness growled and saw them off the premises, in a leisurely fashion. Maria was furious at being made to leave. She wanted to look at the cub – 'such a darling'.

I moved my tent to Selya Lake, on the opposite side to the Vicars-Harris's camp. After Swynnerton had inspected Noel's work, I was asked to join them for tea which we had in front of their tent. Suddenly 'Stevie', Maria's tame vervet monkey, came gambolling out of the tent with crimson lipstick smeared on his black face, and rouge and mascara blotches on his greenish fur. Maria leapt up

19

swearing in some foreign tongue. Swynnerton rocked with mirth: he had farmed near Mozambique, knew Portuguese, and understood her comments on Stevie.

Swynnerton's dictum was that motor transport might be used only if loads had to be carried which were too heavy for a porter. One day I got a message to go into Kondoa next morning to see him. I left at 5.30 a.m.: it was a 9 mile walk, 4 of them through heavy river sand. On arrival, he told me that he was very sorry that Burtt was away for the day, as he wanted me to discuss a problem with him. Swynnerton's latest idea was to see if tsete would die if fed on an ox which had had an arsenical dip. Suddenly he said, 'Oh! you will be walking back now, return this afternoon with some live flies and your loads, then you can talk things over with Burtt this evening, spend the night here, and experiment with dipped cattle tomorrow.' I did 27 miles that day, and on the next day did not succeed in leaving Kondoa until late in the afternoon. I had to do much of the journey by moonlight at a fast speed to escape an approaching storm. I made it, but the poor porters got soaked. Such was Swynnerton's magnetism that one never grumbled.

Shortly afterwards, I had a bad stomach upset. I wished I had some castor oil, remembering my father's advice that if you have eaten something bad, get rid of it. So I walked into Kondoa to see the doctor – a famous character. His sole qualification was a Licentiate of the Apothecaries' Hall, Dublin. I asked him for some castor oil or whatever he advised.

'What– what-what-what, my boy. You don't want castor oil, you want champagne. Nurse, a bottle of champagne from Medical Comforts.' We drank it between us. I repeated my request, and all it produced was a second bottle, so I went to see Emson who produced some veterinary castor oil and made me stay the night. The native's name for the doctor was 'ndege nbaya', literally 'the bad bird', but a name applied to the night-jar because it is held in superstitious dread. It also says 'What– what-what-what'. Some months later I got a bill for two bottles of champagne, because the auditor, very rightly, refused to accept it.

Learning Swahili was difficult as none of my field staff knew any English. I was reliant on a vocabulary compiled by a missionary of delicate sensibilities. On one occasion I wanted to say, 'Take these cows to Nerai village'. I looked up 'take'; it was 'twaa', so I used 'twaa'. The staff rocked with laughter and the person addressed rolled on the ground in his mirth. Later I learnt that 'twaa' is used in the sense of to take a woman.

One day I was exploring a little known area. I emerged from dense thicket into an open space at the foot of a steep-sided, low hill supporting candelabra euphorbia trees, so-called because of their upward pointing branches, and bristling with enormous slabs of rock poised at an angle of about 45 degrees. I climbed the hill and found that the slabs could have provided ideal rock shelters for primitive man, so I started to search for rock paintings. I suddenly saw a daub of what looked like blood. On stepping back, I realised that it was a painting of an elephant with a man walking behind it with an up-lifted stick, like a herdsman behind a cow. Since there is no history of the African elephant ever having been tamed in the past, I was much intrigued. I went on to find paintings of giraffe, and what might be a schematic human figure. These findings formed the basis for the first scientific paper I ever wrote, and led to a visit by Dr. L. S. B. Leakey, the famous Kenyan anthropologist, which will be referred to later.

Subsequently, I used to climb all rocky hills in search of paintings. On one such occasion, an exclamation from Ali, a fly-boy, revealed a large python with its tail two feet from my foot. It glided away, and with great deliberation coiled itself up in a niche under a projecting rock. The fool of a boy, hurled a large stone onto it. Instead of going for him, it came for me. I had a glimpse of a gaping mouth and white teeth before I took to my heels. I ran until I was breathless, and looked round to see the grass bending and swaying some twenty yards behind me. I started off again leaping from rock to rock until I got off the hill. I suspected that the python had been badly hurt, as I could not go fast on such boulder-strewn ground. What I said to Ali is unprintable.

In mid-November, bush fires were at their peak. They are deliberately lighted to make a firebreak around a village and to clean the farmland and adjacent bush, but once started they may go for miles. In a strong wind the flames rise to a remarkable height and the fires sweep through at great speed. Three such fires were converging to meet at Selya Lake, so three of us went down to see if we could intercept any fleeing game. We stood in a large clearing almost surrounded by 14 feet high papyrus. We saw nothing and were about to go home when I shot a guinea-fowl. Immediately the papyrus was converted into a raging sea; the heads were waving, bending and swaying. Then the waves of movement converged, and a huge one

21

swept away from us. All that time we had been standing near a large herd of buffalo, confident in their concealment.

Life became very much easier when Swynnerton very kindly lent me Saidi Abdullah to look after my field staff of 15 fly-boys – a motley group, locally engaged. Saidi was a Game Scout, who had killed a wounded lion with his hunting knife, and had been badly mauled in the process. He wore dark blue putties, a shining brass buffalo-head badge on his pill-box cap which was trimmed with leopard skin, as was the sheath of his knife. He always had his .303 rifle slung on his shoulder. He had been a sergeant in the German army and had been wounded in the war. He was a martinet. Every morning at 6.45 a.m. he would parade the fly-boys in front of me, after a ceremonial salute and clicking of heels. He was illiterate and knew no English, but a genius at interpreting my Swahili. Further he taught me how to hunt and stalk. For work that needed rough timing, I gave him a 5 shilling Ingersoll 'turnip watch'; he grasped the principle that one hour meant letting the 'long leg' of the watch get back to where it started from.

Life also became easier in the tent. I sacked my servant Juma for theft; he had a superbly honest face. I replaced him by Musa Chilwa, a Nyasaland boy, with mission education, yet a Mohammedan! On the day I engaged him I wrote, 'I chose him because he looked as if he would willingly cut my throat and then not have the decency to wash his hands'. Noel, referring to my cook and Musa, said that he had never seen such a villainous pair. They both turned out to be excellent. I had Musa until I transferred to Nigeria five and a half years later. He was a little man, who always wore a red fez, white or khaki shirt and shorts; he was spotlessly clean, had an enquiring turn of mind, and was full of fun.

Saidi Abdullah and Musa Chilwa will be referred to frequently in the following pages.

For weeks the country had been parched, and all vegetation shrivelled up by the great heat. Everywhere there was loose sand, which the wind would periodically pick up and carry for miles, looking from a distance like brown rain as it drove down the valley. Insect life seemed to be scarce; at mid-day the bush seemed to be dead. It appeared impossible for those dead, withered trees to come to life – let alone mile after mile of yellow grass or grey ashes, and black thorn bushes.

22

Then some of the more venturesome trees came into leaf; they seemed to know that the rain was coming. This was confirmed when the rain-birds came and circled two or three times round the lake, but then, having delivered their message of hope, they disappeared and left the bush to go on baking. Life still seemed to be lying in concealment and waiting: few insects came to the lamps: no animal was to be seen in the bush: the country seemed to be depopulated. The atmosphere was charged with suspense.

Then the frogs in the marsh increased their croaking; their note had changed and was no longer so monotonous. They knew, and were excited.

The next day the suspense became intolerable: every living thing was holding its breath and waiting. Early in the morning, the south easterly breeze was filled with the scent of freshly-wetted earth. At mid-day the bush was like a furnace: there was no wind. Every moment one expected the withered grass to burst into flames. Then, on the far horizon, a massing of black clouds appeared; there was a distant rumble of thunder. Soon the sky was overcast, but no rain came. The suspense became intolerable. The cicada's note became shriller and shriller, until it penetrated through one; the normal wheezing note was replaced by a high-pitched, piercing, never ending scream. At dusk there was still no rain, but for the first time for days, the moon was obscured; the frogs croaked more furiously than ever. Then the suspense broke: rain began to fall, then it fell in sheets and a wonderful sweet smell arose from the quenched sand. It rained and rained all night.

On falling asleep, full of the expectation of a lovely cool breeze that would come with the dawn, there was instead an awakenment to a steaming atmosphere with a sickly smell: it was a revolting world. But on going out of the tent, where yesterday there had been a waste of black ashes, today there were tiny little blades of green grass.

I walked down the Saranda road, which hitherto had appeared as a monotonous track of yellow and orange sand, punctuated by funnel-like burrows down whose sides slither the prey of the young ant-lion who is waiting at the bottom. Today there was no sign of them. The wet sand was swarming with ants, some huge and black, some small and red and some yellow.

The road was littered with countless thousands of transparent greyish wings of female termites (white-ants) that had emerged from the ground for a single nuptial flight, after which their wings fall off.

23

A day or two later I went on foot *safari* with Burtt. He was delightfully mad, 6 foot 4 inches in height, very thin, endless knock-kneed legs draped with dirty putties that were all wisps. When he walked his legs sagged. His sky-scraping head was furnished with horn-rimmed spectacles. His back was draped with a haversack fixed by yards of string. He always ate 12 eggs at a sitting. He would meander through the bush with his helmet on the back of his head, and a pair of rose-clippers in his hand with which he would periodically remove twigs of trees for identification; they had endless names. He was a crack shot, loved by everyone, and very eccentric. He gave every passing native a slap on the bottom, said that he was very sorry but had mistaken the projection for the tail of a cow, at which the injured party went into howls of laughter. He was brilliant at his work. Quite a number of names in the East African flora are – *burttii*. I learnt much from him.

We set off with 40 porters who went on ahead. Burtt and I walked through endless bush. We climbed and climbed until we reached 5,300 feet and then dropped down the other side. On reaching Amathle we found that Saidi Abdullah had got our tent pitched; all was ready including two lime juices on the table. Having drunk four each we felt better. It was a lovely site, surrounded by great hills of granite boulders. The tent was pitched under a fine-leaved acacia. The green of the young leaves was very welcome.

The next day we set off for Thlawa. Our local porters belonged to the Wasandawe tribe – a primitive people who speak a 'click' language from the back of the throat. One of them put down his load, leapt into bush, seized an enormous frog – bigger than his hand – and plunged his teeth into it, a nauseating sight.

Burt was busy teaching me plant identification. We stopped at a bush with white flowers, heavily scented.

'What's that?'

'I have no idea.'

'What's the scent remind you of?'

'I don't know.'

'Women, of course!'

It was a gardenia.

By the time we reached Thlawa, our porters were querulous; they had had no meat since we started. Having had breakfast, we set off at dawn next day with a local guide, followed by the blessings of the village headman who was not disinterested in our fortunes.

We walked through parkland, bounded on each side by thicket. After some four miles Saidi Abdullah spotted three eland about a

24

quarter of a mile ahead. Burtt had shot all the eland allowed on his licence, so the responsibility was mine. Saidi started flitting about from bush to bush and I followed behind him. Then a short crawl until the last bush was reached. I took up a sitting position but there was a branch in front of me, so I had to recline on Saidi who steadied me with a fatherly arm. The bull was staring at me and refused to browse, so I cautiously raised my rifle and fired; he continued to stare at me, but Saidi said not to shoot again. The bull trotted a few yards and fell – stone dead.

The porters rushed up like wolves. Soon the hide was off, the entrails out and the middle stomach produced; duly emptied, the porters refreshed themselves on raw tripe. There was lots of meat for everyone; the eland is the largest of all antelope, the bull stands nearly six feet at the shoulders and weighs some 1,200 pounds. I gave the village headman a shoulder; he gave me the equivalent of a handshake, by rubbing the palm of my hand with his little finger. Soon dozens of small fires were burning, encircled by skewered meat.

In bad tsetse areas cattle cannot be kept, the people suffer from meat hunger, and will do anything for meat. In the days I refer to there was abundant wild game, and one's staff expected to be kept supplied with meat. This was the secret of keeping a happy family atmosphere.

On the following morning, I watched natives collecting white-ant females off the trees and eating them alive. Burtt had some fried for breakfast and said that they were very good.

The village headman insisted that he take us round the hamlets under his jurisdiction. He wore a white cap, a khaki shirt with tails hanging outside his khaki shorts and, below the thinnest of brown legs, a large pair of patent leather boots with pale blue socks turned over the tops. We were warmly greeted at each hamlet and sat on stools. A huge gourd of native beer, made from fermented maize, was passed round. I tried it; it was very pungent and tasted like the sourest of sour milk and had lots of things floating about on top. I gave up, but Burtt lapped it up.

As we walked along, Burtt broke up huge lumps of elephant dung, and picked out the seeds. Apparently, a number of tree species are dependent on elephants for their distribution; the seeds fail to germinate unless softened by the elephant's powerful digestive juices. Following behind elephants at this season is hard work: in muddy ground each water-filled footprint may be up to two feet in diameter and ten inches deep.

After a fortnight's enjoyable *safari*, it was good to get back to my own dry tent at Selya Lake, to put on clean dry clothes and sleep in dry bedding. Shifting camp each day, as we did, one gets soaked walking, the folded tent and bedding gets still damper on the porters' heads, and on reaching one's destination the tent is pitched on soaking wet ground.

I went into Kondoa for Christmas where Jock Emson kindly put me up. The European quarter lay on the east side of a tributary of the River Bubu. At this time there were only four scattered compounds, with white, German-built houses, two for the administrative officers, one for the doctor and one for the veterinary officer. The walls were of stone and mortar and the roofs of corrugated iron. The houses were well-raised above ground level – one climbed many steps to reach the front door. The servants' quarters were at the back. The most noteworthy house was the District Officer's – the '*boma*' which had been built with defence in mind; impassable prickly sisal had been planted round it, the only access being by a single path which was covered by a Maxim gun installed in a turret. (These cautionary measures by the Germans were now obsolete.) In addition there was Noel's house – mud bricks, and mud floor surfaced with cow dung and covered with native mats. These compounds were interspersed by huge-girthed baobab★ trees, with smooth grey bark – some with tormented arms raised to heaven in protest at their own grotesqueness.

The African town lay on the other side of the river; there was a main street with thatched mud houses, and a few with corrugated iron roofs incuding an Indian shop – Kangi Damani's. The European houses were usually sited at a distance from African settlements in an attempt to reduce the incidence of insect-born disease, such as malaria and dysentery, among the susceptible whites.

On Christmas Day Emson produced a ham for breakfast: I had not tasted ham or bacon for two months. We spent the day visiting each of the four houses and having drinks, followed by a bachelor dinner at Emson's.

Kondoa was, for Kondoa, packed. An American millionaire and his wife, on a hunting trip, occupied the rest house. Bampfylde, our District Officer (D.O.), was putting up a Mr. and Mrs. Higgins and their two children from Dodoma; on Boxing Day we were all to have dinner with Bampfylde, followed by a fancy dress dance. There would be 12 of us, 10 men and 2 women.

★*Adansonia digitata*

There had always been a joke, without any foundation, that Burtt was married and had a wife in England. It was decided that I had to go as 'Mrs. Burtt'. Mrs. Higgins provided the clothes, including undies: a mauve frock, bare arms, silk stockings and blue bandeau round the head, with a slightly protruding curl. Maria, an expert at make-up, made a brilliant job. I stank like a polecat. I had a fan, powder box, and necklace. At dinner, the servants were deceived, or bemused, and served me first. I flirted violently with Bampfylde, who was dressed as a clergyman and made frequent compromising remarks. All went well until my drawers slipped down whilst dancing. Two people took my garters as keepsakes.

Such was a typical Christmas in a small station. People coming in from bush for a get-together, and making their own simple amusements in the absence of a Club, tennis courts or a golf course.

On the 22nd February 1928, I received a Christmas hamper from my mother, despatched by the Army and Navy Stores. Two tins had broken – how they smelt. We ate the rest on March 2nd, when I threw a Christmas party at Emson's, and so managed to return his hospitality.

Hitherto I felt that, apart from learning the ropes, I was wasting much time and energy. The lower numbers of tsetse available in the country west of Kondoa made it impossible to get experimental results that had any significance. I yearned to get off on my own to a remote area where tsetse swarmed and Swynnerton would forget my existence. Every time I met him he had new ideas of what he wanted me to do. He was like a saucepan whose content never stopped boiling and frequently splashed over. He was always on tour and no one ever knew where to find him.

On getting back to Selya Lake, I started writing up the results of my pre-Christmas *safari*, but before I could report my findings I got a letter from Swynnerton, posted in a distant part of the territory, telling me to do what I had just done. So I dutifully went back again.

The water at Mongoloma was full of mud in suspension and quite thick to stir; the addition of alum made tea possible but not washing, as I had not enough alum to spare. That evening I shot a dikdik, the smallest of the antelope – the size of a hare. I was looking forward to the next night's dinner, but in the morning the cook said that '*dudus*' had eaten the dikdik and a chicken. Since '*dudu*', according to the dictionary, is an insect, and the chicken was a live one, I voiced my disbelief. He took me out to show me the spoor – a feline; it turned out to be a civet cat.

One evening I caught a fire-fly; I was amazed at the light it gave. My lamp was attracting insects by the thousand. Some would certainly get in my soup, so I moved the lamp, and holding the fire-fly in one hand, fed with the other.

Next morning, I saw a long green snake lying across my path; it was busy trying to swallow a chameleon. On my approach it slithered off, leaving its prey behind. I then tied the dead chameleon by its tail to a very slender twig, so that it dangled about four feet above the ground. I found a place to hide, and soon the snake returned. With one or two coils on the ground it then erected its body and tugged at the chameleon, and finally got it down, half-swallowed it, and then slithered away. It was an amazing balancing feat.

Later we came to a spot where elephant had recently passed; the sprouting grass had been brushed to an angle of 45° by their shuffling feet. Shortly afterwards we saw and heard the honey-guide answering honey hunters, who were imitating its call from the other side of the valley. The evolution of the co-operation between bird and man is so extraordinary: the bird calling man to come and open up the tree with the bee's nest, knowing that man will leave the bird some of the comb as a reward. Saidi Abdullah claimed that the honey-guide will also lead man to a wounded buffalo or lion by flitting from tree to tree.

Recently there had been three interesting Court cases in Kondoa.

In one, a European on *safari*, had a lot of money stolen by one of his porters, but the culprit was unknown. There was a witch doctor of repute in the village, who was called in. The porters were lined up and the witch doctor made each hold out a hand, palm upwards, and put a pinch of white powder on it. He then harangued them, and said that the powder in the hand of the thief would become moist, it would then eat into the palm until it reached the bone, and that then the hand would fall off. After further talk, he examined the palms; one was moist, the owner's effects were searched, and the money found.

In the other rather comparable case, each suspect was given a corn seedling and told that if he was the thief, his seedling would grow twice as fast in the night as any one else's. Next morning, examinations of the seedlings revealed one with the head nipped off; its owner was later found out to be the thief. One can picture the culprit, in his superstitious dread of the witch-doctor's powers, watching his seedling during the night, imagining that it was growing at speed, and in desperation nipping off the head.

28

In each case the witch-doctor's approach was psychological; I was to make use of this approach at a later date. My grandfather recorded a comparable method used in India where the suspects were made to swallow uncooked rice – the thief failed to do so.

There was an interesting murder case. A native of Bereku was known to hate his wife so much that his neighbours quite expected him to kill her. His chance came – a leopard entered his compound and took a chicken. He drove it off with his spear, and then knocked his wife down and drove his spear into her. She was dead. He made a few scratches on her stomach and then, in a great state, called his neighbours and related how the leopard had sprung at his wife, knocked her down and was standing over her: how he had rushed to her aid and driven down his spear, but the leopard slipped aside and bounded away. Alas! he had speared his wife. A perfect defence – fresh leopard spoor in the compound and scratches on his wife's stomach – but by chance Jackson, our zoologist, was there and noted that the spoor was large and the claw marks minute. The husband was convicted.

Returning from *safari*, I was approaching Selya at about 6 p.m. when I heard the 'Hoo – Hoo – Hoo' of hunting dogs. I had almost reached my camp, when I heard shouts and yells. The pack had swept into camp and were tearing at the flanks and belly of a young waterbuck, which then fell. The fly-boys had rushed out of their huts with sticks and *pangas* – cutlass-like knives – and driven off the dogs which are the size of an Alsatian, with yellow, black and white patches and white tipped tails; their large ears are bat-like with rounded ends. Soon the fly-boys were cooking waterbuck meat.

I found *Mzee* Hassani – old man Hassan – waiting for me. His crops were being eaten by waterbuck, so I promised to get to his farm by dawn next morning. I had only left camp a few minutes, when I heard the most disgusting snarls. I had walked into a pack of some twenty hunting dogs, which I expected to be miles away. Presumably they had been hunting all night, and by chance had killed near the camp. It was still too dark to see the sights of my rifle, but I fired at a misty shadow. Of course I missed, but to my amazement the pack did not run away. I had been told that they do not attack man, but I did not relish being in this dimly seen, snarling, barking semi-circle of dogs that showed no fear, which seemed all wrong. Every time there was a puff of wind, a filthy stench floated past me. I had two more shots at shadowy beasts, whose behaviour seemed singularly provocative, but to no apparent effect. Suddenly, they all loped off uttering hoarse, half-stifled barks, and *Mzee* Hassani appeared; he had come to look for me. Presumably the dogs had got his scent as they were downwind of him. It turned out that I had killed one of the dogs.

In those days one was asked to kill every hunting dog one could as they were a menace to game, killing more beasts than they could eat, and driving the game out of the area.

Musa Chilwa, my priceless servant, came and asked for a holiday, as he wanted to get married. I gave the bride 3 shillings with which to buy some cloth, and off they went to Kondoa where he got a licence from the *Boma* (District Office) and a ceremony and blessing from Sheik Ali. (I knew that the bride had been living with him for a year and that they had a baby, but Musa had suddenly thought that it would be a good idea to get married.)

Musa came back that evening and reported with glee. I asked him if it was not a strange feeling to be married. He gravely said that it was a very strange feeling. He then told me how he had gone to his father-in-law, and the conversation that had ensued:

'I have had your daughter for a year and have decided to take her.'

'Good my boy. I am glad that you found her satisfactory, but what about the little matter of 160 shillings for her? She is worth it.'

'True *mzee*, but I consider 110 shillings ample. After all, her child was a girl.'

'Yes! Well let us say 120 shillings.'

'Done!' says Musa. 'Here is 20 shillings down, and I will pay you 10 shillings a month.'

Musa then showed me the receipt for the first instalment of 20 shillings and asked me whether his father-in-law had made it out correctly. He then took 10 shillings off me for the next day's wedding feast.

Musa left, and my late cook came into my tent, and asked whether he might spend the night in my camp, as Musa had engaged him to cook the wedding feast. 'A strange country this', I thought, 'but how delightful.'

I was very pleased about it all, as now Musa could not get tired of a life of *safari*, as had the late cook, since it would take him ten months to pay for his wife. In fact, we had become very fond of each other, and I did not think he would ever leave me. He never did.

30

CHAPTER 3

Solo – Without a Road

This was a period in which I saw very few Europeans. Hitherto I had been working in Wasandawe country to the west of Kondoa town. Some 17 miles to the east of the town lies the Masai escarpment, which later runs due north of Kondoa to Ufiome Mountain. Below the escarpment lies the great Masai steppe; in places the escarpment rises to 2,000 feet above the plain.

I had been ordered to look for a suitable site for a research station below the escarpment. I was delighted because tsetse were reported to be very numerous; further, the area was undisturbed by any reclamation work.

Jackson, the zoologist, was to undertake a reconnaisance on top of the escarpment, but we joined forces for the first few days. We left Kondoa in early February 1928. The first night was spent at Kolo, the Sultan's headquarters.

The next day's trek was the worst I ever experienced in 32 years in tropical Africa, even though the distance was only 22 miles. After two hours march we descended the escarpment by a rocky defile, rough going but no trouble. Then we had 12 miles on a wide track along a sandy river bed, enclosed on each side by trees with the midday sun beating down on us. The track followed the foot of the escarpment which rose like a wall; the rocks threw out tremendous heat. It was stifling. The porters should have been two hours ahead of us, but a number had foundered along the path. A nervous impluse made one swallow, with the result that the sides of the throat got stuck together, and only a convulsive cough seemed to clear the blockage. By 2 p.m. we were both exhausted. Jackson decided to push on as he must get a drink at any cost. I decided to wait and let it get cooler – not that it did. After an hour I had to push on: the tiny sweat-bees nearly drove me mad, crawling all over my face and eyes in search of moisture, and the furious probing of numerous tsetse flies was not conducive to rest.

31

On reaching camp at Kandaga, I found Jackson just recovering. I emptied three pots of tea, but in a few minutes I was just as thirsty. I waited half an hour, and them emptied a syphon of soda water. I had frequent drinks during the night and a sore tongue next morning. We were all dehydrated. It was the wet season, and should have been cool but not a spot of rain had fallen for six weeks.

Next day I shot a hartebeest, which cheered up the poor porters, and Jackson left me. I spent the following 10 days exploring the area around Kandaga. The hamlet, consisting of only two little compounds, lay in a small bay in the escarpment.

There was one extraordinary character, called Ishmaeli, who had been in the King's African Rifles. He was a tall, raw-boned, gaunt man, quite black, with European features, who claimed to be one quarter Scot; he knew no English.

Ishmaeli always went to bush with a double-bladed sword, with its scabbard suspended from his shoulder with a strip of black felt, worn as would a Knight of the Bath wear his sash. Another strip of felt suspended from the other shoulder carried his water bottle. He wore a cap on ceremonial occasions, such as escorting me to bush, with a peak like an English railway porter's with bits of tin sewn onto the brim.

Ishmaeli was just like a vulture; as soon as he heard a shot, he would appear from nowhere. On one occasion I was far out in the plains and had just dropped a zebra. I was walking up to it, when Ishmaeli appeared ahead of me. Presumably the zebra was not quite dead, because suddenly the great sword was drawn and the gaunt figure, swinging it over his head, brought it crashing down through the animal's neck. He would do anything for meat, and when I saw him perched on a carcass, eating raw liver, I wondered whether he had been a vulture in some previous existence.

In bush most people wore khaki belted bush shirts, which hang outside the shorts and have buttoned breast and side pockets, the latter being most useful for ammunition. Boots, with pigskin gaiters or puttees, provided protection from thorns and snake bite.

At Kandaga one boot pinched my little toe unbearably, so blaming the boot I inserted cotton wool between toe and leather, but to no avail. After some days of misery, Musa Chilwa casually mentioned that the camp was infested by sand fleas. I knew about jigger fleas being a pest in tropical America, but did not know that they had spread to Africa: the pregnant females usually burrow under the skin of a toe nail. I asked Musa to investigate; he gave a gurgle of delight and told me that there was a monstrous jigger under the nail. After some painless surgery with a needle, he proudly presented me with her ladyship – with tiny head separated from a tiny tail by a circular sac full of eggs. Had he broken the sac, I might have got a very nasty, septic toe. The hole under the toe nail would have held a pea seed, but it healed very quickly. Incidentally, the sailor's exclamation 'I'll be jiggered' stems from this flea.

When a newly arrived South African discovered that his servant had extracted a jigger from his child's toe, he told his wife that in future the doctor must be asked to do it. He did – and the child cried out 'Do let Audu do it; he doesn't hurt'.

Some years later, when I was in Northern Nigeria, the jigger flea reached a primitive hill tribe. They had no idea of what to do, and some lost limbs and some died from secondary infections, before news reached the medical department. A nurse had to be despatched with a large supply of needles to teach the people how to cope.

After ten days at Kandaga, I had to walk back to Kondoa to attend a conference on research, before Swynnerton went home on leave.

It was decided that I should do research at Kandaga and at Kikori – a village 18 miles further north. I was to spend alternate weeks at each place, but since porters would be difficult to find, my weekly trek would have to be done without any. Hence I would need duplicate equipment. A camp was to be built at each place for self and a staff of eight. I was given 75 labourers for eight days only, as by then their month's contract would have expired.

My Kandaga house *cum* office was to be a one roomed native hut of the *'tembe'* type – 12 feet wide, 20 feet long, and 7 feet high. The framework of the walls consisted of forked poles – small trees cut where the branches formed a fork; the poles were inserted into holes dug in the ground, and thin sticks were closely lashed across from pole to pole. Water was poured on earth from white-ant heaps and was well trodden to convert it to mud of a suitable consistency. Balls of mud were then hurled against the framework to which it adhered and the surface smoothed with the hands. The flat roof was similarly constructed. I had three forked tree trunks down the middle of the house to support the roof; nails hammered into them provided pegs for clothes, rifle etc. The flat roof was just a very thick layer of well compacted mud. (Many a native has been killed by the roof collapsing in the rains).

Initially I left the supervision to an *askari* as I thought I would be best employed if I got some meat for the gang. On the second day a labourer came to my tent and complained that the *askari* had beaten

33

him, at least hit him with a stick; he was accompanied by a growling mob. I asked the plaintiff to show me the weal. He bent over, and unwound half his blanket to reveal an unmarked bottom. I gave him a resounding smack with the flat of my hand, and told him to go into Kondoa and tell the magistrate that I had given him a terrible beating. This sally was greeted by howls of laughter by his friends, and the plaintiff was delighted at the thought that I had smacked his bottom and that now we were on intimate terms.

On the third day I worked with the gang whose motto was 'Survival of the Slackest', as we still had to finish my kitchen, a hut for the *askaris* and older fly boys and another for the young ones. At first the labourers were very idle. I was hurling mud at the framework and noticed that one man was picking up the smallest of handfuls, so I pressed a handful of mud on his head. This got everyone laughing, and soon they were chanting and working to rhythm. When my house was completed, I christened it Zebra Hall: it had a mud floor, a doorway but no door, and two window holes but no windows.

Whilst building the Kandaga Camp, I had a sick parade of about seven people a day, two seriously ill. It was very worrying, as I had little knowledge and few drugs. I felt that the Colonial Office ought to have had a short course for the laymen going out to the tropics for the first time, and that one should be given a medicine chest of simple drugs before going to bush. Our Kondoa doctor, who was shortly retiring, gave me rolls of pink boracic lint, cotton wool, iodine, Epsom salts and only one bottle of quinine tablets. When I had to send for more, he sent me a bottle of quinine powder with no instructions. I knew that the dose at that time was 30 grains a day for fever but, when one has no scales, what does 30 grains look like? My rule of thumb had to be:– *all fevers* – quinine and asprin, when available: *all bowel troubles* – Epsom salts.

I was luckier than most, as my father had fitted me out with a small medicine chest, but how pitiful were the drugs available in those days:– tincture of arnica for bruises, Dover's powder to induce sweating, asprin, potassium permanganate for snake bite and as a disinfectant, sodamints for indigestion, cascara, castor oil, zinc oxide ointment, ipecacuanha as an emetic, but best of all Dr. Collis Brown's chlorodyne to alleviate dysentery – one could even buy it in the Indian *duka* in Kondoa. (When I had bad veldt sores, I had to write to my brother, serving in Egypt, who, after consulting the army doctor, sent me an ointment which worked). Fortunately it was not long before we got a most excellent doctor at Kondoa.

It was noticeable in those days how one sometimes got a misfit among the older government officials who were mostly competent and dedicated men. It was said that, during the First World War, having driven the Germans out of a district, the Commanding Officer would leave several people behind to administer the area; he naturally chose the least efficient and when peace came these men, being 'experienced', became members of the Colonial Service.

Having built the Kandaga camp, we moved on to Kikori – some 20 miles to the north. It was a weary march which necessitated scrambling over the hot rocks of the Kisesse pass which crosses an eastward projection of the escarpment; here we saw klipspringers – a tiny antelope – leaping from rock to rock. I often saw them, as I had to make this journey between my two research stations at weekly intervals. (Bicycles were useless, as the long acacia thorns produced endless punctures.)

Kikori was a thriving little village situated in the mouth of a bay in the escarpment wall. I sited my camp on a low tongue of hill, which bisected this bay, and looked down on the village and beyond to the Kisesse promontory and beyond that, some 40 miles to the south-east, to a protrusion of the escarpment with a white-faced cliff which we christened 'the Ghost Mountain'.

Having built a kitchen and staff quarters, I found that funds were almost exhausted, so I decided to use my tent as a bedroom and bathroom and to build myself a hovel for my living room and laboratory. It was about 12 feet long, with a doorway at one end, and 6 feet wide. The mud walls supported a thatched roof with eaves, but so low that I could only stand up under the ridge. Satirically, I called it 'Rhino Hall'. I had to pay off my labour gang before my house was finished, but got it completed most cheerfully by the villagers in return for meat.

My first effort to get them meat was not a success. I dropped a waterbuck with a bullet through the base of the neck, collected tsetse off it, and then held its horns for the Mohammedan guide to cut its throat. Suddenly, my arms were nearly wrenched from their sockets, the beast leapt to its feet, scattered us like chaff and dashed off at great speed. We stood and gaped. Presumably, the bullet had just nicked the spinal cord, causing temporary paralysis. We were following the tracks which led into dense thicket, out the other side, and then doubled back, when the guide stopped and listened and then, looking very frightened, said 'that is a very bad bird'. I listened and heard it saying very quickly 'Cha: Cha: Cha:'. The next moment there was a crashing of branches, the guide shouted *'faru'* – rhino – and shot up a tree, my gun bearer shot up another tree, and I leapt behind a bush, as the rhino shot past like an express train. Fortunately, the rhino disappeared, as I, stupidly, had no hard-nosed bullets on me – they were in a haversack up a tree – nor had I a full game licence, as I could not afford one. However, I had learnt a lot, in particular the alarm call of the *'chasi'* tick-bird or red-billed oxpecker, who was to warn me on many occasions of the presence of a sleeping rhino – a common beast at Kikori. We followed the spoor of the waterbuck for two and a half hours when we lost it, but I do not think that he was badly wounded.

Kikori was a beautiful area. The hills were clothed in *Brachystegia* – lovely feathery-leafed trees which shaded out-cropping slabs of rock; lower down it gave way to *Berlinia* whose leaves are larger and less delicate – pink, then pale green in the first flush, but later darkening. Both closely related genera become leafless in the dry season and form what is known as *miombo* woodland which covers vast areas and was referred to by weary foot-slogging Europeans as MAMBA country – Miles And Miles of Bloody Africa. Down in the plains the *miombo* gave way to many species of acacia some forming belts of woodland, interspersed by areas of dark dense thicket, or open areas of long-grassed *mbuga* – seasonal swamps. After some six miles one reached the short-grassed Masai Steppe, into which the crystal clear Kikori stream emptied itself.

Three of the acacia species in the plains were noteworthy: the small gall acacia, *formicarum*, with long thorns protruding from oak-apple-like galls: the much larger *xanthophloea* with unhealthy-looking green trunks reminiscent of Kipling's 'all set about with fever trees', and as one neared the steppe the wide-spread, tall and flat-topped *spirocarpa* with dark trunks giving a park-like effect with wildbeest and zebra taking the place of deer.

The Kikori people were quite different from the primitive indigenous tribes. They had originally come from Kilosa, some 200 miles to the south east, and had brought with them a knowledge of irrigation; there were acres of sugar cane. Kilosa had been on the slave route, so either the Arabs or the Kilosa people had introduced mangoes, bananas, custard-apples, bread-fruit and even a little coffee. The villagers, some hundred people, were mainly Mohammedans. There was a mosque from which chanting would float up to my camp, very melancholy and like the droning of bees. The houses had thatched gabled roofs, not flat mud ones. The

36

women wore sheets of gaily printed calico tightly rolled above the breasts and hanging down to below their knees. The men, when working wore grey blankets tucked in around the waist, but on important occasions cotton gowns with a white cap or red fez. The villagers and their headman, *Mwanangwa* Tanganeza, were always helpful.

Game was abundant, which was as well, as there were no cattle owing to the swarms of tsetse flies.

Behind my house, and for my own personal use, I had a pit latrine dug, with two forked posts linked by a smooth one to sit on – all surrounded by mats for privacy. Next morning when I went to use it, I was furious to find that someone had been there before me. Everyone denied having done so. Tanganeza who happened to be in the camp offered to bring up his 'doctor' to find the culprit. I accepted, as I wanted to see the method used.

A tubby little man arrived, wearing a red fez and white gown and holding a little red book written in arabic script. He was followed by two, similarly dressed, disciples with a bowl and each holding two sticks, with a blood stained notch at the end. The bowl, filled with hot embers, was placed in front of the 'doctor'. His disciples sat cross-legged in front of him, facing each other, and with the bowl between them. The 'doctor' then started chanting lugubriously from his book and crumbling a resinous substance onto the hot ashes in the bowl which produced a column of smoke. As the tempo of the chanting increased, each disciple raised his two sticks and, with careful aim, notched them into the sticks of the man opposite, leaving a space over the mouth of the pot. Then, to set an example, Tanganeza faced the 'doctor' and lowered his fist into the mouth of the pot and said 'If I used the European's latrine, may Allah expose me'. Every member of my staff had to follow this practice.

Often the sticks slowly closed in on a wrist and then, amidst a sigh of relief from the onlookers, they slowly parted. The suspense was terrific. Then came the turn of Nsoro, a fly-boy with a squint eye which was always in a state of antagonism with its fellow. He finally got his arm rightly aligned, and his fist between the sticks. Slowly the rods came together, inch by inch, until they were almost touching the trembling wrist . . . they snapped together. The 'doctor' with accusing finger, shouted 'I've caught him'. Nsoro's deficient eye revolved rapidly, and twinkled. He had only just arrived back from Kondoa.

37

Having built both camps, I started my field research in March 1928. The subject must be briefly mentioned, as it was to form the background to a number of events.

I wanted to study fluctuations in the size and distribution of the tsetse population in relation to season and the different vegetational communities available. For example, a certain type of vegetation might form a sanctuary for tsetse in the very hot dry season; if so, it could afford a key to the control of the insect. To study this subject I devised a form of 'fly-round', in which two expert fly catchers followed fixed paths, divided into sections according to the vegetation type traversed. All tsetse caught were killed, put into the appropriate tube for that section, and brought back to camp where I counted and categorized the catch, which might be up to 1,600 flies. Observations were also recorded on the game and spoor seen, weather and meteorological conditions, bush fires etc. A fly-round would be done from say Section 1 to 10 on one day and, after a gap of one day, in reverse from 10 to 1. (I was always a catcher on one of these two days). There would then be a ten day interval so as not to disturb the game. Three such fly-rounds were set up: the two at Kikori were each about 7 miles long and were continued for 4½ years, the one at Kandaga was 12 miles long, but was given up after six months.

The tsetse fly is an unusual insect in that the female does not lay eggs but gives birth to a full grown maggot, which rapidly burrows into the ground to form a pupa. We also had pupa-rounds to establish whether the breeding grounds varied with the season.

The shooting of game was not just undertaken for sport and to feed the staff, but also in an attempt to establish the tsetse's most favoured hosts for a blood meal: as soon as an animal dropped, one would rush up to it and catch the tsetse on the carcass, and also collect external and internal parasites for the British Museum of Natural History.

Having laid out the two Kikori fly-rounds, I returned to Kandaga to demarcate a similar round. I was sad to hear that Tony, my white duck, had been taken by a leopard during my absence.

A few nights later I heard the leopard coughing – to my ear it is more like the double note made by the pull and push of a saw. At 6 a.m. Musa brought me my tea and told me that the leopard had taken my salted hartebeest meat, hanging in a muslin bag on the kitchen door-post, despite the fact that he and the cook were sleeping inside. Then Saidi Abdullah arrived and told me that he had followed the leopard's tracks: first it had climbed onto the roof of the partitioned staff hut, in which 12 people were asleep, and had taken my hartebeest head and hidden it in the grass: it then returned, tore down the matting curtain over the entrance to the two young fly-boys' room – they woke up, saw the leopard in the doorway, and screamed, but no one heard them: the leopard had then trotted 30 yards up to the doorless, doorway of Zebra Hall and watched me sleeping through the mosquito netting curtain: it then stole the meat from the kitchen and had carried it down to the opposite wall of my house where it ate it, leaving the bag behind: finally it had carried the hartebeest's head up the escarpment. I quickly dressed, and we followed the tracks up the very steep hill-face but with the tall grass and rocks it was hopeless. I began to wonder whether the leopard's intentions were strictly honourable.

Sitting up for the leopard was impossible as my torch batteries were dead and there was no moon, so Saidi Abdullah made a rifle trap. He built an oval stockade, just large enough to hold a tethered goat, with one small gap between the posts. My rifle was lashed point downwards, with twine attached to the trigger, then up over a typewriter reel, and down to a peg driven into the ground at the centre of the gap. If perfectly set, the leopard is shot through the skull as he pushes through the gap, but if the gap is too narrow he gets a bullet through the nose and if too wide he gets badly grazed, leaving a very dangerous animal in the vicinity. The leopard ignored the trap.

Life at Kandaga was like living in a zoo without bars. The camp was surrounded on three sides by the towering escarpment which was thickly clad with trees interspersed with great slabs of black rock. On the fourth side I looked across to the horizon. In the foreground, a tongue of steppe reached the meagre farmland. On either side bands of well-watered green grass gave way to bands of gold and yellow separated by black bands of dense thicket – all sloping gently down to the distant steppe where there was nothing but grass, apart from a pair of ridiculous pimply hills. To the north, beyond Kikori, I could catch a glimpse of Mt. Ufiome, and to the south-east the chain of hills ran out into the steppe to end in the Ghost Mountain, with 9,000 ft Mt. Mkongo in the background. It was an idyllic spot, so peaceful during the day, so noisy at night when there was a background of whistling, chattering, gurgling, hissing and odd squeaks round my hut and, as I was to find out when the steppe game came in for water – the roaring of lions and the braying of zebra. It was two months since I had seen a European, and I was very comfortable, but it did not last for long.

39

It was Good Friday evening, and being very tired I went to bed at 7.15 p.m. It was raining heavily. I woke up to find myself soaked, as rain, mixed with the white-ant-heap earth of the roof, poured through. Photographs, clothes, boots, papers, office reports and table-cloths were drenched and slimy with mud. I bundled the more perishable articles into a tin uniform box. I then made a cover over the camp-bed, by tying a waterproof groundsheet to the mosquito net rods, having first thrown all the sodden bedding onto the floor. I found one dry blanket in a box, rolled myself up in it, and was soon asleep. Several hours later I woke up with a start, half suffocated, to find streams of water pouring into the bed. The weight of mud and water that had collected in the groundsheet had become so heavy that it had snapped the string. I gave up, put on some dry clothes and a mackintosh, found a drier spot, and sat and read a six-week old newspaper and Punch by the light of a hurricane lamp.

Next morning, I sent one man post-haste to Kondoa to get a tarpaulin and another to the nearest village for 20 labourers to replace the earth on the roof. I went off from 6.30 a.m. to 1.30 p.m. to do the Kandaga fly-round, and was soon wet again. At this season it is not the rain that wets one, as that comes on in the late afternoon, but the dew-soaked, 6 ft high grass, which soaks one from the waist downwards; then the grass dries, and one becomes wet from the waist-upwards with perspiration. Needless to say, I got back to Zebra Hall to find that Musa Chilwa had dried bedding and clothes in the sun, and that all was shipshape.

For several months I had noticed small mounds of earth which I attributed to some small rodent. One day, I heard a noise like distant bagpipes and ran it down to its source. A very large mole-cricket, with his wings raised at an angle of 45°, was stridulating furiously: the noise was deafening and sufficiently piercing to be painful. He was sitting beside a small hole under the lee of a mound of earth. I excavated a number of these mole-cricket burrows. The hole leads down to a 'living room', from which another passage leads up to the surface and opens on to the far side of a second mound, which is about a yard away from the first; it is an escape passage, in case an intruder tries to get in at the front door. From the 'living room' a third passage leads almost vertically downwards to a lavatory – a small chamber filled with mouldy excreta. From this chamber a small passage leads horizontally into a larder containing neatly chopped blades of grass.

I returned to Kikori, as next day I was to have my first visitors. They were walking from Bereku, on the Kondoa–Arusha motor road,

40

down the hills to Kikori. Dr. Phillips – a doctor of science – was a recent appointment; he was to be in charge whilst Swynnerton was away on leave. Phillips was an experienced plant ecologist who had come to check that my fly-round paths had been laid-out in accordance with the vegetational communities traversed. Whereas Swynnerton was a brilliant, if erratic, naturalist, Phillips was a very able scientist who understood my reason for insisting on large numbers of tsetse flies for my research and was to teach me much about the collection of meteorological data etc. He was accompanied by our second botanist, St. Clair Thompson.

Having checked my fly-rounds, Phillips insisted that I should take a couple of days off and accompany them next day on an expedition up the lower slopes of 8,000 ft high *Mt. Ufiome. We camped in the grounds of Galappo Mission, which had been established in 1910, and had become a thriving community before war broke out. In a minor engagement 27 soldiers were killed and are buried in the mission cemetery. One gravestone to 'George Rex' is intriguing. He was a descendant of George Rex of Knysna in South Africa. According to local legend the first George Rex was the son of George III before he came to the throne, and thereafter became an embarrassment to the king. In consequence he was shipped out to the Cape in 1797, held the title of Marshal of the Vice-Admiralty Court and given a Crown grant, – on condition that he never married. However, he did marry and had children, and moved to Knysna where he had a huge estate. Phillips knew the area well and said that the local legend was that George Rex lived in splendour, considered himself to be the true king and displayed with pride miniatures of his father. It was a good story which made that cemetery in such a lonely setting even more poignant.

After our soldiers left, it is said that Sultan Dodo took the corrugated iron off the mission roof, together with timber and anything of value. When I saw the place the grounds were choked with weeds, all buildings roofless, and windows glassless, with one young Father doing all he could to restore the mission. When I returned some years later, miracles had been achieved.

Deneys Reitz, in his book 'Trekking On', gives a good picture of this part of the Territory as it was in 1916. He was with General van Deventer's column which had started in Kenya and whose objective was Dodoma on the Central Line; to the east was General Smuts whose objective was Morogoro, also on the railway. Reitz describes Ufiome as a little village which had been a German administrative

*On modern maps Mt. Kwaraha.

41

post with a substantial *boma* (fort). He moved southwards to Kondoa Irangi and camped in front of the *boma*; at dawn they were bombarded by guns removed from the cruiser Konigsberg – a wreck in the Rufiji delta. Desultory fighting continued whilst both sides waited for the fever season to pass and the roads to dry up. When the southward drive continued, over 300 men were left behind in the Kondoa graveyard. On capturing Dodoma the column moved eastward to join up with Smuts. Shortage of water, deaths from malaria and dysentery and the loss of the horses from *nagana* – make grim reading. Yet only 11 years later there were so few traces of those epic days.

The labourers who built my camps were of the Ufiome tribe. It was good to be amongst them again. They are beautifully built, slim, perfectly formed, with small heads and finely cut features. Both men and women were naked to the waist. The men carried magnificent spears, and wore four or five circlets of copper wire round their necks, and always a necklace of pale blue beads. Later I was to find that members of this tribe on the remote north eastern face of the mountain were very different.

We set off at dawn to explore the mountain with a guide. The lower slopes were clothed with bushy vegetation interspersed with grass; it was nostalgic to see forget-me-nots, scabious and burnet saxifrage, and painful to contact a nettle that stung like an electric shock. Soon we were enveloped in a wet mist. Buffalo spoor was plentiful, but as we climbed it was replaced by elephant tracks, broken trees and flattened grass. It seemed extraordinary that an elephant could keep its footing on so steep a slope.

We entered the rain forest. In places the elephant paths formed tunnels through the dense vegetation. Elsewhere great trees were visible, festooned with beards of grey lichen. The lianas appeared as thickening cables as the eye followed them down from the tree tops. The undergrowth was rank and twining. Everything seemed to be strangling everything else. It may have been due to the mist, but my impression of the rain forest was that it was characterised by premeditated mass-murder by parasites and hangers-on. It was awe inspiring, but I felt ill at ease.

On looking upwards, one often saw the bark of a tree that had been rubbed smooth by an elephant, which made one realise man's pygmy stature; sometimes tree trunks had been deeply scored by elephants sharpening their tusks; everywhere there were piles of dung, providing the dung rolling beetles with a massive problem.

On a steep decline, there were several deep-grooved ditches, indicating where the elephants had slid down.

The air was full of the heavy smell of humus as we wound along dark tunnels, interrupted by lighter glades, when we heard an elephant below us making snorting, gurgling noises as he drank. By the time we could see the pool, he had left. We ate our sandwiches to the accompaniment of elephant snorting and breaking off branches across the valley. We had to return before reaching the top of the mountain, as it was 2 p.m. and the return journey was long.

Soon after the Ufiome trip, Saidi Abdullah and I were doing a Kikori fly-round far out in the steppe, when we saw a large number of vultures. We reached the spot to find a few bits of meat and a pile of dung. Saidi took a quick look, cocked his rifle and said "four, four, four", and crept to an adjacent thicket, ignoring my existence. I followed him. There we found the meatless carcass of a huge bull eland. Saidi was peering first in one direction and then another, and then, seemingly satisfied, straightened himself and said something I could not understand, so presumably he swore. (All this took about two minutes). Then he relaxed, noticed my existence, and an awed Dr. Watson humbly listened to Sherlock Holmes.

We emerged from the thicket. Saidi pointed to the tracks of the eland as he browsed, then showed me the faint footprints of four lions creeping up behind; then to a series of deeply indented prints showing where the lions dug their claws in before springing. Then the churned up ground where a tremendous struggle ensued, the broken sapling, and lastly the feast. How at dawn they had dragged the carcass into the thicket and finished their meal, and then plodded off to drink. He explained how he had hoped that they would eat only half the kill, and would be sleeping beside it. How, being gorged, they would have been content just to see us off and how, when out in the open, we would have killed all four. What a man! (Lion were classified as vermin in those days).

The native ideas of what England must be like are curious. Musa Chilwa came into my hut and asked me how many kings there were in England. I replied "one". He looked incredulous, but after some fathoming I got to the bottom of his trouble. He had been reading an old "Illustrated London News" and with his limited English concluded that we had numerous kings. He brought me the magazine and showed me King Macbeth, King Lear, King Alexander the Great and King Neptune, all of whom he considered to be our reigning monarchs. (The picture of King Macbeth in modern dress might have been a trifle misleading).

Musa was also very taken aback to hear that there were no tsetse flies in England, no lions, no elephants and not even buffalo. It took me five minutes to persuade him that we don't stock lions. He then looked at me in an accusing manner and said "Well! why you put lion heads on all your English things and monies"? I realised that the prestige of England was at stake, that soon it would dawn on him that we were a nation of boasters and liars – a nation which represented itself as the "Country of the Lion", when it could not boast a single indigenous specimen. The honour of England was saved in the nick of time – I had a brainwave. I told him that we never called ourselves "lions", but that the nations of the world were so terrified when we went to war that they said we were not men but "lions". I hastily added "female lions", as the male has a bad reputation for courage. Fortunately there was a tin of Norwegian "Lion Brand" sardines on the table. I snatched it up and pointed to the huge ferocious head depicted on the top, and told him that, though Norway was a far country, when they made food up for England they put a lion's head on the wrapping to show that it was a food fit for lions to eat. Having saved the honour of my country, I had a drink.

On every visit to my Kandaga camp I was intrigued by the extraordinary pre-dawn calls made by some unknown animals. The call sounded like 'Prr! Prr! Prr! Kwa! Kwa! — Kwa! — Kwar! Kwar! — —Kwarr! —— Kwaa! ——— Kwaa!' The last notes were fainter, and sounded much further away than the first. Then another individual would repeat the call from quite a different direction, and then another and another. This saying of "Good morning" would last for half an hour or more, and then cease abruptly. If heard close-to a tremendous volume of sound was produced.

Enquiries amongst my staff and the locals elicited the reply that the noise was made by the deadly crowing-crested-cobra, a very evil beast. Descriptions varied greatly. One version was that the snake is pure white, as long as a python, lives in grass or reeds, feeds on frogs and will attack man with the utmost ferocity: death is speedy unless a 'doctor' of the Warangi tribe is available to give him medicine. I recorded the nicest version which I got from a relative of Ishmaeli. "Abdullah, my father, killed the snake and saw its cock's comb. Truly he was a brave man, my father. Often he would tell the tale of how he passed along the path frequented by this terrible snake, with a bowl of steaming porridge on his head: how the snake had suddenly shot its great head out of its hole, high up in a dead tree, and had struck at his head: how the jaws plunged into the scalding porridge and the snake dropped onto the path, writhing and twisting in its last death agony".

One day I solved the mystery. At first light, I heard the call coming from the tree by my hut. I rushed out with a gun, and shot the vaguely seen beast. It was a poor little tree hyrax, a grey, guinea-pig-like animal some 16 inches long, whose relative the rock hyrax (rock-rabbit) lived in the boulders above Zebra Hall and used to whistle at me. Needless to say, none of the Africans would agree that the animal shot was the maker of the dreadful call. Incidentally, the coney of the Bible refers to the rock hyrax.

I had never suffered from boils, but on this visit to Kandaga I had a nasty feeling that I was developing a crop of them between my shoulder blades. Having only a little shaving mirror I could not examine the area. Periodically, I would get a short, sharp pain, just as if a red hot needle had been placed on the flesh. I was sitting in my camp bath, when Musa came into my hut and must have looked at my back. He said "Why don't you let me squeeze the maggots out of your back; they are just ready?" It all came back to me – that lecture on myiasis in man: how the African *tumbu* fly sometimes lays its eggs on cloth that has a body odour: on hatching, the maggot burrows into the skin: when full-grown the "boil" develops a black head, and the maggot can be squeezed out. Musa proudly presented me with five of them.

After seven weeks of peace, one of the two Kandaga camp guards arrived at Kikori to say that they were again being terrorized by the leopard. On the first night, he had entered their hut despite the hurricane lamp which I had given them; one of them had woken up to find the leopard standing beside him, tugging at his tunic on which he was half lying. He yelled with fright. The leopard seized his tunic, and bolted; next morning it was found in my lavatory torn to shreds. At 10 a.m. that morning the leopard entered Ishmaeli's compound and carried off a goat. That night the leopard again raided my camp, but was foiled by a door which the guards had made for their hut.

I returned to Kandaga and for five nights Saidi Abdullah and I sat up for that mad leopard. We sat with our rifles resting on the mud

*Cordylobia anthropophaga

window-sill of my windowless window, and a torch for which I had now got batteries; the torch was attached to my rifle. The bait was a kid tethered outside the hut, with its mother inside with us; there was much bleating.

It is an ordeal to sit up night after night. There is nothing to focus the attention, but one has to fight sleep. I would do mental arithmetic, and imagine that my brain was working vigorously, my head would drop forwards, and I would experience the sensation of falling. I would then renounce sums and concentrate on the kid – the way his tail stuck out, what a huge swollen stomach he had, how absurd were his little horns – and then I would realise that I had already made those observations one hundred times in the last hour. Worse still, when the kid's head fell forwards and he went to sleep – then the real test came. The night noises helped: the "what! what! what!" of the night jar and Kondoa doctor: a distant lion grunting, but best of all the "Prr! prr! kwar!" of the tree hyrax indicating that night was ending, as did the cold early morning breeze flowing in from the great plain, and finally a cock crowing.

Twice we heard the leopard in the distance, but he never came to the bait: once we thought he had – a black form slipping through the grass, the kid mute and trembling, a flash of the torch and a hyaena bolted away.

During this "sitting up" period I got a wooden door made for Zebra Hall by two village experts from the plateau, whose sole tool was an axe; they had never heard of hinges, yet made me a door that swung on an axis. The principle is worth recording, and is best explained by an analogy. Picture two straight-stemmed tobacco pipes laid horizontally on a table, one say six inches below the other, with the stems on the left and the bowls on the right. The bowl of the top pipe faces downwards, and that of the lower pipe upwards. They are united by a pencil sharpened at each end (the pivot post) which is fitted into the bowl of each pipe (the sockets). Secure say a post card to the pencil and you have a swinging door if the contraption is raised into the vertical position. The 'pipes' are made by the very careful selection of two small trees each having a *straight* branch growing out at a *right-angle* to the trunk. Each branch is cut off at some six inches from the trunk and a hole burnt into its cut face to form a socket. The trunks of the two trees will be cut off at say one foot to the right of the branch and say three feet to the left. One will be mounted horizontally as a lintel, supported by two forked poles, one on each side of the doorway, and the other placed on the ground as a threshold with the hollowed branches carefully aligned in the vertical plane. A small stone is put into each socket as a "ball-bearing" before the first post is inserted. The door itself is made of two horizontal poles, lashed to the pivot post, with vertical rough-hewn planks between them.

Of course, if the leopard was intent on getting in he could do so through the window holes in the mud walls, so I kept lamps and bottles in them, so that he would have to announce his entry. (Incidentally, pressure lamps had not yet come in, so one had to try and read by the feeble light of hurricane lamps).

At the end of this visit we set off for Kikori, leaving Musa to sweep out the house; having finished, he followed us. Before long he heard a leopard grunting, looked round and saw it, with drooping head, walking parallel with the path at some ten yards distance. Musa kept his head and walked faster. On hearing our chattering, the leopard disappeared. Then Musa joined us and did the chattering.

It may well have been atypical, but in the Kondoa area at this time the administrative officers invariably sided with the African. I sent two apparently cast-iron cases into court, concerning theft of Government property, but they were both dismissed. This made my ex-German *askaris* very contemptuous of British rule, but not of the British. They much preferred German rule because they knew exactly where they stood. For example, a German coffee grower was murdered by his labourers in Mbulu district. His body was found buried in a coffee pit. When Saidi Abdullah heard the news, he looked very disgusted. I pointed out that five of the six murderers had been caught and would doubtless hang. He replied "What is the use of that? In German times every man in the village would have been shot, so that not even one would have escaped". A very fierce outlook by our standards, but one that the African of those days seemed to appreciate.

Food was always a problem. For breakfast I had porridge and tinned milk followed by fried eggs, but no bacon; for a change I would have fried antelope brains. For lunch in bush I would take sandwiches containing tinned Indian butter and minced antelope meat which was always tough as it went bad if one tried to hang it. For dinner I had chicken, or stew, or fried antelope liver. (Once I thought I would have a change and have kidneys in the stew, but on the first occasion I used the wrong word and got gall-bladder instead). If an antelope had any fat on it, the fat would congeal on the roof of the mouth, whence I removed it with a dessert-spoon. I would celebrate birthdays with a tin of bully beef. The pudding course was always rice or tapioca. Fruit was very welcome when in season. The water always had to be boiled and filtered.

I tried making biltong – strips of meat, salted and hung up in the sun to dry, but was off-put by the swarms of bluebottles that laid their eggs on it, and its high flavour. (I have always loathed anything high as a result of stories of an ancestor who could not get venison in Ireland, so he buried linen-wrapped legs of mutton in the garden. Later in life the blowfly was to become my greatest ally: when asked out to Sunday lunch in Nigeria, if the servant bringing in the usual roast duck was followed by a swarm of flies I knew that the bird was high, tipped Wendy the wink, and we both took small helpings).

In most departments the field staff moved into a head-quarters town for the heavy rains, but this is impossible when one is doing field research. I was delighted when the dry season started in mid-May, as for three weeks I had had no flour, my boots had worn out – a pair only lasted me for three months – and my feet were growing fungi as they had been wet day after day. In June, Phillips insisted that I should go into Kondoa for three nights – I suspect that he thought that I was beginning to talk to myself. Getting in was now easier as Noel Vicars-Harris was making a motorable track, running along the foot of the escarpment, and Phillips had no inhibitions about the use of motor vehicles. The making of a motorable track consisted in cutting down the bush, burning out the stumps and filling in holes with earth, or stones if available; streams were bridged with straighter tree trunks, criss-crossed with branches and surfaced with earth. Such roads were adequate in the dry season, but terrible in the heavy rains.

In Kondoa I replenished my provisions, managed to buy a pair of South African boots called "Prime Movers" from the Indian store, and got the new doctor to repair my feet. But the crowning event was that Maria gave birth to a baby boy, much to Noel's delight.

Soon after my return to Kikori, I shot a magnificent greater kudu with 52 inches long, spirally, twisted horns. I first saw him about a quarter of a mile away down in a valley. I left the fly-boys and I and one *askari* stalked him for one and a half hours through eight feet high grass which necessitated the climbing of a tree at frequent intervals to see where he had got to. (Such 'elephant' grass consists of slender canes, with occasional side leaves, topped by a thin head of seeds; when dry it burns furiously). My suspicion that this antelope is a very important host of tsetse was confirmed when we took 61 flies off the carcass. On its flanks were the old claw marks of a lion that had failed to make his kill. We were lucky not to have been charged by a rhino as unwittingly we had passed within ten yards of one, but down-wind, as was discovered by the fly-boys when they caught up with us.

One Sunday, I heard a tremendous noise in the camp and was wondering whose wife was getting beaten, when there were shrieks of "The *bwana*! The *bwana*" and a scurrying of feet. I found all the *askaris* and their wives outside and the spokesman solemnly asked "Does a snake drink water?" I said "Yes! it drinks dew." There were wild shrieks, reminding me of the declaration of a poll – a scurry of feet as they rushed back to their quarters and shouts of "Father says that a snake *does* drink water." I then overheard a member of the opposition say "Allah! he only knows about tsetse flies."

An hour later the Kikori Debating Society reappeared. I was asked whether it was true that there is no such thing as a female hyaena, as the locals never caught females in their traps. They said that when a male hyaena wants a young one, he takes a piece of meat in his mouth and runs some miles with it. He then digs a hole and buries it. Some days later he returns, opens the hole, and finds a young pup waiting for him. Having overheard the opposition's rude remark about my ability, I improvised rapidly and replied that when a pair of hyaenas went hunting, the male always led the way, and so was always the one to be caught in the trap. This explanation appealed to them greatly, and there was no argument. (It was 43 years before I fully understood the reason for this apparently ridiculous local belief. In the Van Lawick-Goodalls' fascinating book 'Innocent Killers', the authors describe in detail the visual similarity between the genital organs of the male and female hyaena. A trapper asked to bring in six hyaenas, three of each sex, soon produced three 'males' one of which produced triplets whilst he was still searching for females).

On the same day, a third deputation arrived. The leader said that all of them knew that the white-breasted crow had no blood in its body, except for two fools whom they could not convince. I had to shoot a crow to convince them that the fools were right.

On another occasion I was reading a six week old Sunday Times, when I heard much shouting and saw two parties coming up the hill yelling at each other. When they got near I saw that one was led by Saidi Abdullah and the other by the *askari* Musa Mauridi. On arriving, both saluted and clicked their heels, and Saidi began

49

"I found this man sleeping with my wife. I want him flogged."

"Is this true, Musa?"

"Yes! *bwana*."

"Well! you must both go into Kondoa, and you Saidi must bring a charge against Musa in the courts."

"And get him fined 10 shillings for sleeping with my wife. No! *bwana*, I want him whipped with a *kiboko*."

"What about you Musa?"

"I don't want to walk to Kondoa and back. Whip me *bwana*."

"How many strokes am I to give?"

"Twenty strokes," said Saidi.

"Twenty for that woman. She was not worth one."

Shouts and yells. Followers restrain patrons.

After much bargaining, it was agreed that I was to administer six strokes with a bamboo cane. Musa bends over, and I start to administer the punishment to the accompaniment of "Harder, *bwana*, harder," from Saidi.

After the six strokes, Musa straightens up and flings me a salute as does Saidi. Both walk off, arm in arm, the best of friends, with their supporters intermingling and chattering away. Honour had been satisfied.

I decided that it was time I shot my first rhino. We left camp at 4.30 a.m., and after half an hour's walk reached the edge of the rhino country, and waited for dawn. On moving on we soon came upon a large *Kigelia* – the sausage-tree, so called because of its huge sausage-shaped fruits each suspended on a long stalk. The fruit are very hard-skinned, and much loved by rhino; they had been eating them only a few hours before, exposing the luffa sponge-like content. Having traversed a shrubbery we entered parkland with 7 feet high grass. Despite the frequent climbing of trees it was not until we were within about 100 yards that we saw two grey humps with tick birds on them. Saidi Abdullah and I got a bit closer to check that they were males, when up sprang two more – the females. They went running up and down, squealing like giant pigs, trying to locate us. Then the females charged us from opposite directions. I fired one shot over one female's head and she rushed squealing back to the males. Saidi Abdullah said "Climb *bwana*, climb. Trouble for certain." The only tree was a very spiky candelabra euphorbia, but it might have been a staircase from the way we got up it. The remaining female was guarding one flank and the other three were approaching us from the front. As a milky-juiced euphorbia was not the best of vantage points, I felt that I had better disperse the party. I aimed for one of the male's heads, as that was all I could see in the long grass. I got him slap on the side of the head, but instead of falling, he came head on with blood streaming down his face, then he turned sideways thinking that that was the direction of the danger. I guessed how much down in the grass his heart would be, and fired. Down he crashed, and I was just congratulating myself when up he leapt and charged off accompanied by one female. The other two made off in another direction.

We climbed down and followed the wounded beast; the blood was full of bubbles indicating that I had missed the heart and hit the lungs. Every thirty yards we climbed a tree to see if he was lying in wait for us. Soon we came to a little thicket and climbed a small tree to see if he was inside. I heard a rustle in the grass below, looked down, and saw a large male rhino, who was following his wounded fellow, and the female. Two were bad enough, but to follow three was too much, so I yelled. He stopped, and looked up at me at about four feet distance. The shock was too great for him, and he slowly trotted back. (Had it been a female, the reaction might have been different). We followed the tracks through tall grass and dense thicket for a fearsome one and a half hours, expecting to be charged at any moment. By this time we were worn out, and I decided to resume the hunt next day by which time we might find the male dead and that the female had gone off on her own.

I set off early next day with the local guide; unfortunately Saidi Abdullah had had to go into Kondoa for a court case. We took up the trail where we had left it; it was soon joined by the spoor of a third rhino. Then followed four hours of extreme tension. The trail led through numerous thickets and tall grass, came round in a circle on itself, and finally led to the notorious Kikori rhino thicket, about 2½ miles long by ¾ mile in depth – a great belt of thorny shrubs growing under the sickly-green trunked fever trees (*Acacia verugera*). Having reached the edge, Saidi, my local guide, usually a brave enough man, refused to enter the thicket so I went on alone. The thicket was criss-crossed with rhino paths in the form of tunnels some 5 to 6 feet high, and very dark; there was rhino dung everywhere, much of it only a few hours old. I began to realise that I could easily get lost in this labyrinth and that if two of the rhino charged I would have no hope, as the smooth-trunked, high-branching fever trees were unclimbable. I now realised why Saidi refused to enter this home of the rhino, and rejoined him, but I felt very badly at leaving a wounded beast that could still be alive.

51

However, I had learnt one thing – my Manlicher .375 had not the striking force needed for thick-skinned animals, even though the bore was adequate.

A rhino has two interesting habits. It tends to dung daily in the same spot and, having done so, to break up the dung by kicking backwards with its hind legs. The natives have a charming legend to account for the latter habit.

Allah created all the animals and the last to be made was the rhino, by which time Allah was very weary. Rhino trotted off, but all the other newly-created animals rocked with laughter at this absurd beast. Poor rhino felt very embarrassed, and trotted off to drink from a crystal clear pond where he saw his own reflection; he then understood. He trotted back to Allah, went down on his knees and said "Allah! all the other animals mock me, they laugh at me, I beg you to give me a nice fitting skin like the elephant." Allah looked at him. "Yes! rhino, I was very tired when I made you, and I am still very tired, but here is a needle. Go, sew yourself up as you want." Rhino thanked Allah, and trotted off. Then he saw a luscious euphorbia, tore off a branch and ate it. He suddenly realised that he had swallowed the needle. He rushed back to Allah, fell on his knees, and begged for another needle. But Allah was tired and angry and replied, "You stupid, stupid beast, from now on, you and all your kind shall break your dung and look for the needle."

As one who has been a staunch conservationist for many years, I appreciate how wrong this shooting of game must appear to younger generations. It was wrong, but there were mitigating circumstances. How was a young man, who had recently been playing rugby football and tennis, to interest himself when stuck out alone in bush? How, when game abounded, was he to get onto friendly terms with his meat-hungry African staff if he ignored their craving for meat? How was he to get onto friendly terms with the villagers if he ignored their plea for help against marauders? Then there was another aspect. It was the done thing to try and get a record head whose measurement would be worthy of inclusion in Rowland Ward's 'Records of Big Game'. One's best heads were scrupulously cleaned of meat, a process that was completed by putting them into an ant heap. The heads were then bleached in the sun, the horns polished and the trophies proudly hung up in the house to be admired by one's friends. There was a story to each one. In the days before wireless and television, and when bundles of newspapers would arrive that were at least six weeks out of date, game was the principal topic of conversation. How poor old Bill had wounded an elephant which had picked him up with his trunk and thrown him onto the top of a thorn tree etc.

Ionides, in his book "A Hunter's Story", wrote "I know many conservationists, dedicated men, to whom money has meant very little in a career devoted to wild life. Without exception they all began as hunters. Hunting is a stage in a man's evolution if he is absorbed in wild life. He goes through it and usually grows out of it. He observes and he learns on the way, and if he thinks at all his conclusions form his philosophy of life".

It took some time to gain the confidence of Kandaga's few inhabitants. On arrival I had been told that there were no rock paintings, but after five months a local led me to a huge overhanging rock choked with thicket. On cutting our way in we found a vast shelter, but all the rock paintings, except one of a zebra, had been ruined by the superimposition of prints of hands dipped in lime. The floor of the shelter was littered with pottery and the bones of wild animals, including elephant which had abounded on the escarpment until killed by the Arab slave-raiders and ivory hunters. The cave was situated on a promontory a quarter of a mile from my camp. Sometimes awful groaning noises came out of the hill which I attributed to wind, but the natives called it "the voice of the hill".

Whilst sitting in the cave, Saidi Abdullah told me about the *Maji – Maji* (water – water) rebellion against the Germans. The Sultan of Tabora's witch doctor had announced that if the people rose and refused to pay hut tax, his magic would turn the Germans' bullets into water. They rose and there was much slaughter. Saidi's father, a small local chief, being fond of the District Officer at Kilosa, had gone to his assistance, and had been later rewarded. This explained Saidi's admiration for the Germans, the severity of his outlook, and his gift of leadership.

On returning to camp one afternoon, I found that a shortcut had been cleared to my hut and a signboard erected "To Zebra Hall – Private Rode". It transpired that Musa Chilwa had seen the mystic words "Private Road" on a signpost to Government House, Zomba in Nyasaland – his home town. He did not know what the words meant, but obviously they must add to the importance of the occupant. (I think I must have gone up in his esteem since I replaced my pith Bombay Bowler which had turned to pulp in the rains, by a hard Wolsey helmet, complete with the brass, giraffe-head badge of Tanganyika Territory).

Musa had been very disappointed when I had not returned with a rhino. He had no idea what they looked like. This seemed very odd until one realised that in the absence of zoos an urban African would have had no chance of seeing a rhino.

By the end of June the leaves were falling from the trees, the grass was going yellow and the game from the Masai steppe was moving in for water. A young man, newly arrived from Durban, spent a night with me. He did not sleep well. My zoo was in terrific form. The mad leopard of Kandaga coughed all night, the lions were roaring, the night jars "what, whatting", the zebras braying, with the tree hyrax terminating the nocturnal concert.

CHAPTER 4

Solo – The Road is Through

In July Noel's road reached Kikori. The village was buzzing with excitement. Seven Europeans, in two cars, had spent the week-end there. Maria, with the new baby, Michael, was in great demand by the women; they were intrigued by the mass of clothes the baby wore. The children were thrilled by the box on four legs, that barked before it started, and then made a noise like a bee. We had a shoot on the Saturday. Saidi Abdullah was very keen that I should not disgrace him; fortunately, I had a good day and dropped two eland with three shots; I gave one beast to the village, so that they could get the party spirit. It was a wonderful evening, sitting round the camp fire and drinking lager beer whilst amazing yarns were circulated.

There were two rather nice stories about His Excellency, Sir Donald Cameron, the Governor. The first took place at a big dinner party in Government House. When the servant brought round a dish of fish cakes, the most senior guest, either because he was engrossed in what H.E. was saying or was bad at arithmetic, helped himself to two. On reaching the most junior guest there were none left, so the servant went back to the first-served and said '*Rudisha moja*' (put one back). The second story was about Swynnerton who wanted to see H.E. on some matter. He heard that H.E.'s special train was due to stop at some wayside station, so Swynnerton sent his lorry on to the next stop up the line to wait for him, boarded the train, and entered H.E.'s coach. After the interview Sir Donald was reported to have said that 'he didn't know which was the biggest curse in Africa, Mr. Swynnerton or the tsetse fly'. We were delighted because our 'old Swyn' would stick at nothing to get his Department funds or whatever was needed.

Now the road was open I got visitors. I still could not get into Kondoa, because I could not afford a car as I was saving to get married, but I no longer felt cut off should an emergency arise.

A few nights later I had a ridiculous experience. I was asleep in my tent when I heard a loud booming noise. I sat up. The noise stopped. 'Imagination! – too much dinner'. I lay down, and was just getting to sleep, when a persistent booming, moaning noise, startled me. I sat up. The noise stopped. Clearly, I was being watched. Wide awake now, I crept out of bed, drew on my mosquito boots, and seized my rifle. I then remembered that my torch was under my pillow; I rested one arm on the head of my pillow, while I groped for it. The noise started. It was the unmistakable note of an irate and imprisoned bee.

Food was a great problem at Kandaga, so I made a vegetable garden. Carrots, onions, potatoes, turnips, cabbages, dwarf beans, vegetable marrows and water melons all did very well. I had sited my garden near a native rice crop on the principle that zebras prefer rice. I surrounded my garden with a thorn fence, which only the eland broke into, and when the locusts came Musa drove them off. I also bought 18 hens and two cocks for 3/6 from the Wa-ufiome. They slept under the cook's bed, and laid partridge-sized eggs in a broken-down hut. They were funny little pepper-and-salt coloured birds. When Emson dropped in for an hour, he, the importer of prize English hens which never laid, was disconcerted to hear that I had got 61 eggs in 8 days. When my back was turned, he slipped six of my eggs under my camp-bed mattress. Fortunately the faithful Musa, who never trusted the 'doctor of cows' a yard, discovered the trap in time. The camp was wild with delight when Musa spread the tale, and next day Saidi Abdullah said that 'the doctor of cows was as fond of play as a young goat' – a remark that was relayed to Kondoa.

I would often overhear conversations that were going on round the camp fire. One of the funniest was Kompania needling Saidi Abdullah. Kompania had been working for Charlie Goss, a notorious elephant hunter and poacher, at the same time that Saidi Abdullah was taking part in the Game Department's efforts to catch Goss red-handed. He was eventually caught and deported from the country.

According to Kompania, Goss was frequently drunk and would never risk shooting elephant until he was 'soaked'. When drunk he treated his staff very roughly, but they would do anything for him. They made amazing marches, with Goss in a hammock, when the Game Department's staff were close on their heels. His elephant gun was called 'Fatima'. Kompania did not know a word of English, but was a very good mimic. It was very funny to hear him say in a sort of thick, drink-sodden voice '*Pa – mimi – Fat – i – ma*' (give me Fatima), followed by 'Bang'. Saidi would keep referring to Kompania as 'you poacher,' to which the latter would reply 'Well! you may be able to catch lions, but you could not catch us,' whereupon Saidi would spit explosively into the fire and murmur sulkily 'Am I a catcher of mice?'

I was asked to investigate a rumour that buffalo had carried tsetse into a zone 6,000 feet above sea level in the wild country on the northern side of Ufiome Mountain – an area populated by only a few Wa-ufiome and their cattle.

We were standing on top of one of the highest hills at about sunset, when some hyaenas started to wail most mournfully in a valley a thousand feet below us. I noticed that my *askaris* looked ill at ease and very frightened, and one muttered 'the hyaenas of the Wa-ufiome.'

Next day when the sun was hot and all ghostly matters forgotten, I asked Saidi Abdullah what the trouble had been. He replied that to the white people it would seem 'mere talk of foolishness', but that he would try and explain. – The women in that area belonging to a certain sect were in the habit of poisoning their husbands if they became rich. An indispensable ingredient of the poison was a part of a brain, or of flesh from an arm, of a recent corpse. An evil woman, addicted to such vices, invariably kept a tame hyaena which would answer to her call. As soon as the news of a death was reported, such women from miles around would prepare for the evening. After dark they would call up their hyaenas, mount on their backs, and get carried, often for long distances, until their beasts had found where the corpse had been buried. Dismounting, they would strike the ground with a stick, when the corpse would emerge from the gaping earth. Often a number of these hags would arrive on the scene and each having removed sufficient for her needs the hyaenas would be repaid with a meal. Thence homewards, the night's work was done, and the essential ingredient obtained for a fatal brew.

There was much more that Saidi Abdullah told me, such as how a powerful witch doctor might arrive at the scene, threaten the hags with exposure and receive as much as 400 shillings for refraining from doing so: but I have written enough to indicate how spirit-ridden the area was, as I was to return there later. However, that evening there was much subdued excitement in the camp when a hyaena laughed most gruesomely. A very old man had recently died.

The old man was reputed to have been 90 years old. He had been a famous elephant hunter before even the Arab slave-raiders had arrived in that area; he had killed many an elephant with his poisoned spear alone and had won many a heavy pair of tusks. The Arabs had sold him a muzzle-loading gun and then, just when trade was at its best, the Germans arrived and stopped the massacre of elephants. That old man must have seen a lot. He would have known the chaotic days of war between Sultan and Sultan, when the traveller had to present the local ruler with an elephant tusk before he could pass through his land unmolested. Then came the ghastly years of Arab slave-raiding when whole villages were marched down to the coast – yet the land was enriched by the introduction of fruit trees and sugar cane. Then came the ruthless war between the Germans and the local Sultans and Arabs, and finally the recent war when the villagers fled into bush lest they be conscripted as porters.

Next day we returned home. The *askaris* were delighted to leave that spirit-ridden land of witchcraft.

Whilst doing fly-rounds and having my lunch under a tree, I had ample opportunity of watching game.

The reddish hartebeest, from the Dutch 'hard beast' because he is so difficult to kill, is a most ungainly animal with his long face and rectangular horns. I always felt that the less wary animals ought to reward him in some fashion for mounting guard over them. The zebra is a good example. In the heat of the dry season one may see them sleeping under a clump of trees, eyes closed, heads drooping, with ever-whisking tails, waving with clockwork regularity. If it were not for the tails, they would be quite invisible at a distance. The shimmering, quivering heat softens the crisp outline of the black stripes which appear as tree-trunks, with the white skin merging into the bleached grass which stretches away behind the trees.

Should the sentinel hartebeest be feeling sleepy, he will just stand and watch the approaching enemy. At first his curiosity overrides all fear, and in his genuine interest his face looks almost intelligent. One can see him weighing the discomfort of moving with the fear of danger. Then his nerve snaps. He snorts loudly and canters off, with those ludicrous short legs working quite out of time, and his white rump bobbing up and down. At the sound of the snort the zebra thunder away, swaying from side to side, but never fouling each other, and raising a cloud of dust to add to the effect of speed. Wheeling and swerving in magnificent fashion they soon leave old knock-knees and his herd behind, for him to pause to renew his curiosity.

An even more efficient sentinel is the giraffe with his periscope waving over the tops of small trees; the game browse near him and trust him implicitly, as at the slightest suggestion of danger he will lumber off with his family and attendant tick-birds.

In July, the nights at Kandaga became noisier than usual. The plain's game had had to come in for water, followed by the lions who often drove game down onto the waiting females with a shattering vocal accompaniment. The first time I heard it, I awoke to a roar with the note going up, up, up to a tremendous high pitch followed suddenly by a deep bass roar as all the air was expelled from the lungs. This was repeated time after time, presumably by different males, followed by a galloping of hooves. What happened then, I do not know, but next morning we found the remains of a zebra some 300 yards from the camp. The locals said that there was a pride of sixteen lions in the neighbourhood; we certainly saw eight lionesses on the fly-round, bounding away at some distance. They were beautiful to watch; one stopped and peered round the edge of a tree trunk, just like a cat.

The rhino, that ill-tempered, cantankerous old man, could be a nuisance on fly-rounds. On one occasion, when my gunbearer had dropped behind to relieve himself and I was whirling a psychrometer – a dry and wet bulb thermometer in a rattle – a rhino came hurtling out of a small thicket. My reactions being slow, I did the right thing by not jumping until the last second. He shot past. We froze. He stopped about 20 yards away, by which time he was upwind of us, snorted, tested the wind, and trotted off wondering what had woken him up. One section of the fly-round entailed sampling rhino thicket. I was always very pleased to get out of it; a fly net in the hand was no replacement for a rifle.

Saidi Abdullah never trusted warthog, because in the war his platoon commander shot one which charged Saidi's section and broke a man's leg. When I wounded one which came at me, I was amazed to hear him flick off his safety catch, which was most unusual as he was a very cool customer. I doubt if they ever charge deliberately. If one shoots a warthog that is head-on to you, its normal reflex brings it rushing towards you. It made me laugh to see them entering their burrows tail first and wriggling in backwards. On one ocassion I looked back to see Mohammedu *mtoto*, a very fat little fly-boy, with hands on hips, looking down into a burrow; the owner shot out and winded him. This became a camp legend.

Swynnerton had come back from leave and was going to visit me at Kikori, so I put up a large notice 'To Rhino Hall' where the path to my camp left the road. Soon after arrving he asked me, as I hoped he would, why I called my residence 'Rhino Hall'. I replied 'Rhino' because they are so numerous, Sir, and 'Hall' because my house cum laboratory is so palatial. He rocked with mirth and said that he would look into the matter, which he did. In due course it was decided that when funds were available I should have a mud-brick house and laboratory with corrugated iron roof, and that then I should give up Kandaga and have a European assistant and an African clerk at Kikori. This was good news as I had been doing three men's work for far too long.

Soon after Swynnerton's visit, I heard a tremendous noise one evening, with wild chanting coming up from the village. Musa told me that it was 'The dance of the devil'. Apparently, a few months before one of the villagers had become a raving lunatic, and since then had been chained to a large log. That night his sympathetic friends were trying to cure him by a very ancient remedy that had survived the introduction of Mohammedanism.

The lunatic is released inside a circle of men and women. The zebra-hide drums are beaten slowly at first and everyone starts dancing round in a circle, including the lunatic. The tempo increases in speed, and all go whirling round faster and faster. On and on they go for hours, until finally the lunatic collapses on the ground. If the treatment is successful, the evil spirit will leave the man, considering that his host is too restive for comfort; if it fails, then the man is chained up for a year, before the treatment is repeated. I cannot vouch for any of this, as I did not feel that it was my business to go and see what was going on, but certainly the noises fitted in with what I was told.

One morning I had a very unpleasant job to perform. At dawn a native came into the camp, armed with a spear, to say that they had caught a leopard by its foreleg in a man-trap, and that as the beast was very fierce would I go down and kill it. Saidi Abdullah and I found a group of rather terrified villagers, armed with spears, and from the long grass some 20 yards behind them came a constant series of furious rumblings and growlings. We approached. The leopard sprang from the grass to be wrenched back by some three feet of chain. Backwards and forwards, tugging and tearing, snarling and shrieking, with bared teeth and everted claws – its very malignity was such as to deprive one of pity. We each fired a shot and I added a third. We found that the leopard had chewed off three inches of its foot in its efforts to escape: at this stage I was filled with admiration for its courage. According to Saidi this eating of the trapped foot is normal behaviour; having virtually freed themselves they are liable to wait until the owner arrives and then take revenge.

Whilst I was visiting Kandaga, Dalton visited Kikori in the hopes of shooting a rhino. He took my guide Saidi with him and asked him why the Kandaga leopard was so eccentric and cunning, and why I had failed to kill it. Saidi replied 'Oh! that is quite simple, but don't tell our *bwana*. An *askari* whom he got rid of was very annoyed and asked his half-brother of the Wagogo tribe, who is a clever witch doctor, to trouble the *bwana*. This half-brother drinks the sap of a certain tree and at once takes the form of a leopard; he then goes and plays tricks in the *bwana's* camp, but he will never let the *bwana* see him, because he knows that the *bwana* will fire and that when the bullet strikes him he will cease to be a leopard, but will lie on the ground as his own self – a wounded man. He would never dream of trying to kill the *bwana*, but just wants to annoy and pester him. Lately this man has been away from Kandaga, so the leopard has not been heard recently.'

Swynnerton wanted me to look for a concentration of a certain species of tsetse in the Masai steppe, so in early September I set off from Kikori to follow the first part of the old Galappo-Arusha cattle and caravan route. We did 20 miles or more in blazing sun through miles of leafless thorn scrub carpeted with bleached, tinder-dry grass. Uphill and downhill the straight track went, every part of it an exact replica of the part behind and the part to come. To avoid the glare one kept one's eyes a foot beyond one's toes and tramped on regardless of time and distance, wrapped in one's own thoughts. It was too hot to talk and the only interruption to the shiffle-shuffle of sandals on the red dust was the occasional 'Al-lah', as an ingenious tsetse discovered a particularly sensitive nerve centre on a porter.

Suddenly I woke up, as my gaze rested on a mincing machine by the side of the track; on it was written 'Alexander werk'. Then I saw the bones and skulls of dozens of camels, and then cartridge cases, especially Belgian ones. To this day I don't know the explanation of these finds. Probably there had been an engagement in the war, some 12 years before. I believe that camels were brought over from India for transport purposes, without it being realised that they would soon die of *nagana*.

Eventually we reached our proposed camping ground at a spot called Misoui, to find that the water had dried up and only wet mud was left. The 'black cotton' soil yielded no water when we dug a hole in it. The nearest water was said to be eight miles further on, but we had no energy to go further. I gave my *askaris* a bottle of very sour milk, and dined off a tin of peaches; I drained the juice to the dregs. About a mile away the lions were driving game, and a stampede outside the camp suggested that there might be more lion quite near.

61

(Little did I know that my father died in England that day, a fact that I was not to learn for several weeks. The cable reached Dodoma next day, but never reached Kondoa – such were commmunications.)

A water-hole was found next morning by following game tracks. A layer of elephant and rhino dung floated on the surface; when skimmed off it left a whisky-coloured fluid which could be used for tea. The porter's morale was restored when I shot a couple of bull wildebeest. They are amazing creatures with their brindled sides, beautiful tails, shaggy manes, grotesquely large forequarters sloping down to diminutive hindquarters, and horns like miniature buffalo. By way of contrast, their neighbours, Grant's and Thompson's gazelles were singularly beautiful.

This was uninhabited country. We had met no one since leaving Galappo, but Misoui had its disadvantages. Flies swarmed, and at night the tent was black with them. A much worse pest were the mites, that abounded in the grass and crawled all over one's body, penetrating the skin and causing intense irritation. I added Jeyes Fluid to my bath to counteract the smell of the water and possibly to incommode the mites, and kept the lamp at a distance so that I could not see the colour of the liquid. Whirlwinds were another nuisance as they would toss everything about in the tent; on the first night, one overturned my lamp, the spilt paraffin ignited on the floor, but I soon put the flames out with sand. I decided that Zebra Hall, Kandaga, was a better residential site.

Soon after returning from Misoui, I went to Kandaga. At dawn a child came up to my hut to say that a large leopard had been caught by the leg in their trap and that as father was away would I come and kill it? The same circumstances had arisen as at Kikori a month before. Saidi Abdullah had been recalled, so I took Kompania with me. We arrived on the scene to see a large dark coloured leopard, leaping and tearing, and quite mad with rage. The grass was very long, and frequently obstructed my view. I fired and the beast became demented with fury. I fired again and again and could see that I was hitting the shoulder region every time. I suddenly thought "could the natives be right – was this a normal leopard?" However, I felt that if I did not do something effective fairly quickly, the leopard probably would. I fired again – the leopard only got madder. Another shot – the leopard fell and lay still. Kompania, looking greatly relieved, made as if to approach; I just pulled him back in time as the leopard sprang straight for him. Luckily, the trap held and the leopard was tugged back in mid-air. This time I put a bullet into the base of the neck. He fell and did not move. I covered him with my rifle and got the onlookers to throw stones and sticks at the body. At last satisfied, I inspected the corpse. Of a total of ten rounds fired, there were eight entry holes grouped in the neighbourhood of the heart, and on the emergent side a wound of about five inches in diameter due to the expanding bullets used.* The lethal shot had smashed the vertebrae of the neck. I had had one miss.

Before my arrival, the leopard had become crazed with rage, as it had partly disembowelled itself on a pointed stick that held the trap and also had a six-inch slit on the under side of the neck. I have given these unpleasant details of a most unsporting but unavoidable killing to indicate the fantastic vitality of an enraged leopard. I am told that under similar circumstances a lion's behaviour is comparatively placid.

On the assumption that this was the leopard that had given so much trouble, I tried to find a cause for its eccentric behaviour. The stomach was full of grass and twigs; however, its claws were not worn down, its teeth were good, there was nothing visibly wrong with its eyes, and most unexpectedly it was clothed with layers of fat! Although the lungs and tissues adjacent to the heart were pulped, the heart itself was untouched and was only half the size of that of the Kikori leopard: the onlookers all commented on its smallness.

The local inhabitants were greatly relieved that the beast was dead and were convinced that this was the animal that had been terrifying them for so long; certainly there were no further incidents during the next four years. I still have the skull.

Having had to shoot two leopards caught by a leg in a spring trap, I appreciated the bravery of my one-armed old friend Ali Kombo of Kikori. Some years before my arrival, a leopard was caught in similar fashion. Three times a party, armed with spears, approached the leopard but then ran away. So Ali Kombo picked up a club, beat the animal to death, and in doing so lost his arm.

A few days after returning to Kikori – that long, hot tramp – I decided to explore the village stream which loses itself in the Masai steppe. In the cold dawn with its sickly light, the burnt bush appeared singularly depressing – a forest of leafless, charred and blackened thorn trees, rising from a powdery carpet of black ash. The numerous tracks in the ash revealed the many and varied species of game that had emigrated from the steppe in search of water.

*Had I known, a shotgun loaded with S.S.G. would have been better.

We came to a veritable oasis where the stream was choked with tall reeds and grasses, where Saidi, the guide, said that rhino loved to guzzle, wallow and sleep. The wind was in the wrong direction and we realised that our scent would be carried down to any rhino that might be there as we skirted the oasis. Sure enough there was a bang and crash as if two iron plates had fallen together, a smashing of branches and bushes and out shot a rhino and her calf some 20 yards behind us and stopped motionless, undetermined where to find the owners of that loathsome scent – man. I whispered to my *askaris* "quick run" and bolted towards the river so as to get downwind, as the last thing I wanted was to shoot a cow with calf. On looking round I found everyone was with me, except for Kompania who was beckoning me to go back and dashing after the rhino who had started to trot away. I had to run after him and tell him to come back or he would get the sack. Reluctantly he returned and was staggered to hear that we had seen the calf following its mother. In his excitement he had never noticed the little fellow following its parent's tail. I then realised why he had been transferred from the Game Department for shooting an elephant in the Selous reserve. I was to find that he went quite mad with blood-lust and wanted to kill everything he saw.

After this incident, we continued cautiously along the stream, lest we be taken unawares by rhino, when an *askari* whispered "*simba*" – he had just caught a glimpse of a disappearing lion. The *askari* pointed to a fresh zebra kill some 50 yards away, with the rest of the herd grazing peacefully knowing that the lion was now satisfied. Hoping that he might return, I looked round for a tree as I felt that if we were all above normal eye-level there would be more chance of his returning to his kill. The nearest climbable tree was about 100 yards away – a most unpleasant thorn; I slipped off my boots and quietly went up it, forgetting all about the thorns.

Soon the vultures came flocking round and a small grey jackal carefully trotted up to the kill. The more daring vultures lurched onto the ground and stood a few yards away from the jackal. All the partakers in the orgy seemed to wear a guilty nervous look. The jackal snatched a piece of meat, and ran away to eat it at a safer distance. The vultures crowded round in a circle, like gamblers round a card table, and wearing the same look of greed. Often the ugly meat-snatchers cast apprehensive looks over their shoulders. I felt hopeful. The minutes seemed hours and pins and needles crept up my right foot, up my thigh and almost to my waist; then my foot began to tremble and nothing could stop it. Suddenly a magnificent maned lion bounded into view, scattered the vultures, and stood over his kill. I aimed oh! so carefully, but all my right side was trembling. I took the first pull off the trigger and squeezed. The lion jumped into the air, and was out of sight in a second – a clean miss.

I waited in the vague hope that he would return, but I had noticed that he was so gorged that his belly nearly touched the ground, so I was not optimistic. For half an hour, I watched the vultures gulping down meat, then staggering off a few paces, appearing to be very ill indeed, and then returning for more. Backwards and forwards the jackal went, carrying off his piece of meat each time, and then the feast was over. Those vultures that were light enough to rise flopped up into a tree, croaked, and rested for the good of their digestions. (How bitterly disappointed I was at the time and now, in retrospect and in old age, how delighted I am that I missed, yet had the opportunity to observe the sequence of events described).

We followed the lion's tracks for some time. They led to the stream where he had slaked his thirst, then through the ashes of a bush-fire, but we lost the spoor in unburnt country. It was noon and very hot so I decided to rest and go home when it got cooler. We had been drowsing under a large tree for some time, when there was a tremendous snort from some beast that had got our wind. The guide went up a tree but could see nothing, so we continued to laze through the heat of the early afternoon. Finally, we got up and had only gone a few yards down the path to the stream when we found that our own tracks had been obliterated by those of a large rhino.

The return home was uneventful apart from a herd of wildebeest wending their way to water. It was open country so we sat motionless, downwind of them. Slowly and surely they came, plodding along in single file, led by a very old bull. The leader passed within twenty yards of us: one by one they filed by with their heavy heads hanging down wearily. There were 23 of them in single file following the winding track of the leader. So orderly was their purposeful progress that one expected to see a little naked boy following in their wake, driving his tired herd home after the fatigue of a long day's grazing – instead there was a huge bull shepherding the cows and calves. It was a beautiful sight: we did not disturb them.

Musa Chilwa was honest but had strong business instincts. One day I had been climbing a rocky hill with very dense vegetation at the bottom which had necessitated going on my hands and knees to push through it. Suddenly my face started to burn and I realised it had been brushed by the strongly urticating hairs which grow on the pods of the 'buffalo bean' creeper. By the time I got back to the tent my face was on fire, and drew much comment from the porters who were lolling about outside; on looking in my mirror I saw that my face was the colour of a beetroot.

Having quenched my thirst and rested, I called for Musa to bring my bath water. I put a few drops of Jeyes Fluid into it and had a very leisurely bath, and washed my head. By the time I had dressed and entered the front of the tent, my face had ceased to burn and the redness had almost disappeared. Meanwhile Musa had carried out my bath tub. Whilst having my lunch, I heard some jabbering going on behind the tent, and went to investigate. There was Musa selling my bathwater as medicine at sixpence per cigarette tin.

The following story illustrates how much there is to learn when big game hunting and how, without a white hunter or a very reliable African such as Saidi Abdullah, the beginner can make horrible mistakes.

One day I was out hunting with only Saidi, my local guide, in whom I had great confidence. He pointed to some tall shrubs and green grass, an oasis in the middle of a fire blackened waste, and said "there is a rhino or some big animal there, see how the tick-birds are fluttering". Having manoeuvred so as to approach upwind, we reached the edge of the green grass. Saidi climbed a tree to look into the luxuriant vegetation. Soon he slid down and said "*bwana*, there is a magnificent male rhino and a smaller, dark coloured female – the male has a huge horn".

I got to within about 20 yards of the pair, but could only see the female as the male was lying down in the long grass. We very cautiously retreated and approached on a different line which enabled me to study the lie of the land. I suspected that when the sun rose higher the pair would leave their oasis and make for the reeds and thicket of the stream-bed, which would necessitate their crossing a small burnt gap in the tall grass which would give me an excellent view of them. By standing on a fallen tree trunk I could see the grey back of the standing female about 30 yards away. For half an hour I watched the birds busy at their de-tickling work, and fluttering up when they irritated her and she flicked her tail and flapped her ears. She was restless and moved about whilst browsing. The wind was veering round, but retreat was impossible as she was now far too near to me. Just as I was getting desperate the huge male lumbered up with the tick-birds fluttering and chirping in their excitement.

The smaller female led the way and crossed the burnt gap. Then came the male and Saidi whispered "a very fine horn". I fired. The male travelled a few yards and dropped. Then the female went champing round, looking in all directions for the enemy and testing the wind with dilated fleshy nostrils and ears spread out to catch any sound. For some minutes I stood motionless and did not dare to re-load as the last thing I wanted was to have to shoot her. Then when her back was turned we slipped away and climbed trees. We shouted and yelled and I fired shots, over her head – all to no avail. She just got madder and madder. Finally I had an idea and told Saidi to slip down his tree, make a detour and then return on the upwind side and climb another tree, whilst I covered him with my rifle. It worked. She soon got his wind, and with a snort rushed away to the stream.

I climbed down from my tree and went up to view my magnificent, first rhino; but the horn was disappointing. I lit a cigarette and was studying the beast, when to my horror I saw that it was a female. I was furious with Saidi. He was very upset and said that the smaller animal was far too big to be her calf, and pointed out that my rhino had not calved for a very long time; further, that the behaviour we had watched was not that of mother and calf as the smaller animal did not follow the tail of the larger when they moved off.

Later, I managed to get a copy of Blayney Percival's "A Game Ranger's Note Book" in which he wrote "The young one runs with the dam till it is almost as big as she is". I suspect that this was the explanation.

Rhino in this area had a bad reputation. In the previous year a villager had been killed by one near my Kandaga camp; it had horned him as he tried to climb a tree. A similar fate had befallen a European at Pienaar's Heights near Galappo. Shortly before my arrival at Kikori two rhino wandered into the village one night, became alarmed, and charged the mud-built mosque, which they partially demolished.

One day I saw a male ostrich, some 50 yards away, coming straight towards me. The bird trailed its wings as if they were broken, and then seeing me, started off slowly at right angles to its original direction. One wing would flap spasmodically and then the ostrich would sit down, looking backwards to see how near I was. As I approached, the bird would stagger to its feet, lumber on a few yards, and then sink down. Once it appeared to be too weak to stand up, and toppled over on its side, where it lay panting. Feeling certain that the ostrich had been mauled by a lion, I called for my rifle in order to shoot it, when one of the boys called out: "The bird is well and is only acting; it must have a nest nearby". Leaving one boy to watch the bird, the rest of us retraced our steps, and found the nest about 40 yards away from where I had first seen the bird.

The nest consisted of a clearing in the grass, with eight young birds sitting on the powdered black earth amidst broken egg-shells. Other young had scuttled off into the long grass. In addition, there was one egg in the nest. Without touching anything, we concealed ourselves at some distance. Almost immediately we saw the male trotting back in perfect health, accompanied by the female. She must have been away feeding, having left the male to look after the nest – and very well he did it.

I sent an account of this incident to Professor E. W. MacBride, my old zoology teacher. He was so interested to hear of such intelligent behaviour in so primitive a bird, that he sent the account to "Nature" – the scientific journal.

In late October 1928, I went into Kondoa for three nights to attend a meeting of the tsetse research team. During my stay the Duke of Gloucester spent a night at the rest camp, refusing to sleep at the District Officer's house as he was on an unofficial hunting trip. He asked Barnes, the D.O., and Thompson, the A.D.O. and their wives to come and have dinner with him. They all found him a most charming person. The Duke was very short of crockery so everything had to be washed up between courses. He had just shot an oryx, having followed it for eight hours, with horns almost three inches longer than the record and so was in high spirits. He was a great sportsman and although he had seen seven lionesses he refused to shoot one. He was very unofficial and dined in a pullover.

After dinner they all went up to the Barnes' house, put on the gramophone and held a cabaret in which the Duke took part. Tony, their little son, woke up and called out: "I want to see the Prince". He was sleeping in their bedroom which was in an awful mess, so Mrs. Barnes exclaimed: "you can't go in there". "Oh! can't I just," replied Henry, and rushing in tickled the small boy to his great delight. Next morning Tony had his photograph taken between Teare, the huge game ranger and the Duke who was also a big man.

Barnes, who was a bit of a wag, charged the Duke 10 shillings for the use of the rest camp, as only government servants could use it free. He pointed out that the Duke could not possibly be called a "Government Servant", and so the Duke laughingly paid up.

The natives were all very disappointed that the Duke looked just like a European and wore no red robes. They referred to him as "Mtoto ya Sultani" – the son of the Sultan.

In early November, Mr. Tschope, accompanied by his wife, arrived unexpectedly at Kikori to build my house. Initially accommodation was difficult. I naturally gave them Rhino Hall I – thatched roof, mud walls, 6 feet wide and 12 feet long – and cluttered up with my Kandaga effects as I had just closed that camp. I kept my tent, and as a sitting room put up a shelter, roofed with skins to keep off the sun. I then started to build two grass huts, but the first rains fell, the natives naturally wanted to farm and could only be induced to stay by my shooting zebra for them.

Rhino Hall II was to have a corrugated iron roof, sun-dried mud brick walls, three rooms each 12 x 12 feet to provide a bedroom, sitting room and laboratory cum office, and a 4 feet wide verandah with bathroom at one end. The laboratory was to have a cement floor and real plaster ceiling, and other rooms sand floors and bamboo ceilings. After 14 months I was to have a house instead of a hut, and above all a roof that did not leak. It was all too good to be true. The first 5,000 bricks were destroyed by a thunderstorm, but thereafter the building went very well. It cost £200.

The Tschopes were a delightful couple but as I did not speak German and his English was not fluent we sometimes had to lapse into Swahili. Nevertheless I think that the following stories, recorded by me at the time, are in main accurate and worth recounting.

Tschope was a stubby little man and it was quite difficult to get him to talk, but once started he was fascinating to listen to. Before the 1914/18 war he was a naval stoker, then for the first two years of the war he was chauffeur to Colonel von Lettow-Vorbeck, the great commander of the German forces in the East African Campaign. When petrol ran out, Tschope – a Lance Corporal – was transferred to a company. He was the sole European survivor by the end of a year, and then commanded his company for 9 months. By the end of the war he held the Iron Cross, 1st and 2nd class, as well as receiving a copy of von Lettow's book, which was only given to 200 men, most of whom were officers. (Noel Vicars-Harris tells me that in the early 1960's, when von Lettow visited Tanganyika Territory, he expressed a wish to see Tschope who was flown down to Dar es Salaam, by special plane, to meet him).

Tschope's hero was von Lettow who was a very strong disciplinarian. On one occasion it came to his knowledge that one of his officers had behaved with consistent cowardice. When the officer entered his room, von Lettow handed him his revolver and said "Tell your boy to bring this back to me when you have finished with it". Everyone dreaded an interview with Von Lettow; he had an eye that never wavered but seemed to bore through the culprit. Only a few of his intimates realised that the terrible eye was made of glass. Strong man though he was, von Lettow never failed to admit when he was wrong: when boots were in short supply he ordered his staff to walk bare-footed and set the example, but after a few hours he rescinded the order.

One of Tschope's most graphic stories was when he reached the banks of the Ruaha river, hot and sweaty. He tore his clothes off and was washing himself when an English patrol opened fire from the opposite bank. Tschope, stark naked, ran back to the tall papyrus grass where he lay hidden and began to think –

Zen I zink to myself, "How can a man live in zis country without zee sun helmet?" So I run to zee bank for it, and zee bullets zey go ping! ping! I run back with it.

Zen I think, "So good boots, where go I find more?" So I run for zem and back, as the bullets zey go ping! ping!

I lie in zee papyrus for a long time and I zink, "How can a man live in zee zorny bush without zee pantaloons?" So I run for zem, but zee English zey shout, "Take your pantaloons Jerry. We no fire."

I have often heard that this was the last campaign in which acts of chivalry frequently occurred.

In July 1915, the German cruiser Königsberg was finally destroyed by our navy in the delta of the Rufiji river where she still lies – a partially submerged hulk. Von Lettow was desperately short of money with which to pay his *askaris* and sent a small party to the wreck to retrieve some shell cases from which he had money minted. I think it was Tschope who gave me my 20 Heller piece: on the obverse is inscribed – 1916 DOA – for Deutsch Ost. Afrika. The brass coin is crudely made, the edges are rough to the touch and it has a tiny hole through which light can be seen. I remember Saidi Abdullah going into Kondoa, some 10 years after the war, to receive his back-pay from a German official; he was deeply touched that their debt to him had been remembered.

My servant, Musa Chilwa, had a great respect for Mrs. Tschope and was always trying to impress her with his efficiency. One day he proudly showed her a beautifully washed and ironed pair of my khaki shorts which had two holes worn in the seat; these he had patched with a scarlet material. She praised him for his beautiful needlework. Kompania claimed that the two red eyes in my bottom brought us luck when out hunting, on which occasions I was perforce to wear them.

Within 100 yards of my new house there was a pile of stones. The villagers told me that long ago a little house had been built there for a

70

German, but he never lived in it. Further enquiries revealed that in about 1890 a German called Kater had had the house built for himself and a garden planted, but was never to see the produce as he was murdered by the Wa-ufiome two days after he occupied the building. This may account for the little wild tomatoes that grew in the small area around my camp.

Soon after my house was built, I met Tschope in Kondoa. I commented on the smashed window in the cab of his lorry. He said, "Zis morning I turn corner on bush road and see a huge pyton (python) lying across it. I drive over tail, and head he come up and smash zee window. What did I do? I put foot on zee gas and I go like zee bloody-hell."

One morning I was doing a fly-round with Kompania in the rocky hills near Kikori. He was just in front of me and I was about to catch a tsetse off his back when luckily it flew down and settled on the calf of his leg. I then saw the beautifully camouflaged fat coil of a somnolent puff-adder near my toe, lying in a semi-circle with its head just in front of Kompania who was engrossed in catching a tsetse on the front of his shoulder. I said, "Don't move. Listen. Don't move. You are standing in the coil of a snake. I will step backwards very slowly. When I say 'Now', you will look down. Then you very, very slowly lift your left leg and step sideways across the snake and then sideways again, very slowly. I will also step back again. You will then join me." He was, on this occasion, imperturbable and did as I told him. I then called for my gun and shot the brute. I have always loathed puff-adders. They are sluggish, fat and flabby, up to 4½ feet in length with heavy curved fangs that may be 1½ inches long. (I always suspected that if bitten by one it might fail to let go). Like all vipers, the venom causes internal haemorrhage. When alarmed they make a fearsome hissing noise – hence their name.

On November 23rd, after a long day in bush I was having a cup of tea when a car drove up. Faul, the driver, handed me a letter from Swynnerton saying that I was to come in at once: the Prince of Wales had invited the officers of Kondoa to a "sundowner" at 6 p.m. I was wanted to give the Prince news about the Kikori rhino. We had one and a half hours in which to get there, and the road was awful. Faul had no idea of what kit was to be worn, so I bundled everything I had into a suitcase. We set off.

After ten minutes the engine stopped – the petrol pipe was blocked. This happened three times, and then a puncture had to be repaired as there was no spare wheel. A few minutes later when we were going fast on a good stretch, the wheel rim flew off and whirled into bush and the inner tube came out like a balloon. It was 6.30 p.m. and we had only done 18 miles. In desperation I dropped the fly-boy, who I was taking with me, at a village as he had a squint eye. The remedy was efficacious; there was no further incident.

On arriving at Kondoa I went to Swynnerton's house, and to my relief heard that the Prince had been late in arriving and that we were not due to go round to the District Officer's house until 9.30 p.m., by which time they would have finished dinner.

I need not have worried as to what to wear, as the Prince was very informally dressed. He was only interested in dangerous game and badly wanted to shoot a rhino. It was arranged that I should go back to Kikori to locate and study the individual movement of our beasts. The party broke up at 3 a.m. by which time I was very weary.

Early next morning I returned to Kikori, and that afternoon started my rhino survey. I found the fresh spoor of seven beasts and heard that that morning two natives had been charged by two bulls and a cow and had to spend a long time up a tree. The Prince was camped at Ndereda, some 20 miles away. I sent a runner with a letter reporting my initial findings to Denys Finch-Hatton, the Prince's white-hunter. He replied that I should expect them at about 2 p.m. on the 27th.

I spent the next two days trying to find the rhino with the best horn and sorely missed Saidi Abdullah, and Kompania who had gone on leave. I was in the unenviable position of being unable to shoot a rhino if charged, as I had expended my licence, and further would be the laughing stock of the Territory if, whilst trying to tie a red label onto the tail of a royal rhino, I perforce had to shoot it myself. However, all went well. I spotted a fine beast and discovered his dunging place and drinking site.

The village was buzzing with excitement. Wild rumours were circulating. I was asked whether the king's son would distribute much money among the folk gathered to welcome him, and kindred questions.

The big day came. The camping site had been cleared, pots of water had been brought up, fire-wood collected, my *askaris* in newly washed tunics and the villagers in their finery – all was ready by 2 p.m. But no one arrived.

Daily we waited in expectancy, but no one came. It was not until four days later that I heard that shortly before the Prince was due to come to Kikori, he had heard that the King was dangerously ill, had had to set off for Dodoma and Dar es Salaam and sailed home on H.M.S. Enterprise.

72

CHAPTER 5

Two is Company

Hitherto I had not had a European assistant to help in my research or to take over if I was away for any reason, nor an African clerk to do the office work and type reports. My efforts to teach myself to type with two slowly moving fingers had not been a success: in a single report I found that I had spelt 'wildebeest' in three different ways.

In December 1928, Captain V. A. C. Findlay, subsequently referred to as Victor, was posted to me as an "Observer". Whereas he was a man of the world, I was an innocent abroad. Despite a discrepancy in our ages, we got on well. One night he would drink whisky with me and the next night I would drink Bristol Cream sherry with him. (I preferred it when he drank with me). Victor had a car which made life very much easier. Initially he lived in a tent behind Rhino Hall II which was near completion. Tschope had cut a stencil out of a petrol tin and had made a frieze of grey rhinos on a whitewash background, trotting along the top of the three walls of the verandah and finishing at a grey pool of water; it made an excellent background to show off my best antelope heads.

It was just as well that Victor arrived when he did as I had started having bouts of ill-health. An abscess on the root of a tooth necessitated my going all the way to Dodoma, and thence by train to Dar es Salaam, which was unbelievably hot and humid, just to get the tooth taken out. It was good to get back home. (I noticed that the months I had spent alone had made me very self-conscious. I saw some people on the Dodoma station platform whom I knew; instead of greeting them, I scurried away feeling that I must look very odd).

The Kikori camp was often invaded by *siafu*, the driver ants. They move in a procession within a little earth trench which they build along the surface of the ground, often for long distances. The walls of the trench are guarded by the larger, well-armed, soldier ants, whilst the other categories scurry along the track, all going in the

73

same direction. Should you tread on one of these processions where it crosses a foot path, in no time you are nipped round the knees and up your shorts by the soldiers who have sunk their jaws into your flesh. (This can be embarrassing when bird-shooting in mixed company.)

If driver ants invade your home, you vacate it until they have left. They make an excellent job of vermin destruction and also of corpse removal – such as the catch from one day's fly-round waiting to be counted. If a procession is seen approaching the camp, it can be thwarted by encircling the camp with ash. These ants are *said* to drive an elephant mad if a few get up its trunk, and also to be used by certain tribes as stitches for small wounds. (The edges are drawn together, a soldier ant is accurately applied, and having sunk in his jaws the body is ripped off – but I cannot vouch for this).

One night I was rudely awoken by Victor charging down to my house, blaspheming – tearing off his pyjamas and yelling for me to de-ant him. He was a somewhat portly man and every time I pulled an ant off his back, a small drop of blood oozed out. His tent had been invaded and the ants had found their way through the mosquito net and into his bed. I put him up for the night.

For some time I had had a pet vervet monkey like Maria's. I named him Tumbo, the Swahili for stomach as that seemed to be his main centre of interest, apart from biting my fingers: but I had misjudged him – he was merely lonely. I then bought a 3 month old puppy and named him Mbweha, meaning a 'jackal', because he was badly bred and I hoped might have some immunity to tsetse-carried *nagana* as he came from a fly-infested area. They became great friends, apart from whose food was whose, and the monkey would spend hours grooming the dog and eating his fleas and ticks. His great game was to leap onto the dog's back and ride him like a jockey, round and round in circles. I trained the dog to follow game spoor and to point, but he showed no discrimination for the latter: when going through rhino thicket he would suddenly point, I would flick off the safety catch of my rifle, only to find that he had seen a monkey up a tree. The pets became a talking point in camp and Musa loved to regale us with their latest escapades.

A few days after I got back from Dar es Salaam I started my first attack of amoebic dysentery, and went into Kondoa just before Christmas. Dr. Wilkin, the most excellent and kindly of doctors, put me to bed in a tent near his house and that of Maria and Noel, so I was well looked after. Dr. Wilkin had been in the Air Force during the

74

war, was taken prisoner and was the expert lock-picker for the Escaping Club. (He did so dislike being asked to pick the lock of the District Officer's safe, when it could not be opened).

One day I lay in bed watching swarm after swarm of locusts passing. Whenever this happened the D.O. had to notify the Director of Agriculture immediately. On one such occasion, the following correspondence took place by priority telegrams:

D.O.   Large swarm of locusts from north-east going in south-westerly direction X Took three hours to pass X Usual action taken X.

D. of A. What action have you taken?

D.O.   None, as usual.

Late on New Year's Eve a very cheerful party came to my tent to help Nash see the new year in. The Wilkin's flirtatious nanny perched on my camp bed. When the party was breaking up, Dallas, with whom I had travelled out on the boat, disappeared, cut the tent ropes, leaving self and nanny enveloped.

Having had only half the prescribed course of Emetine injections, I had to return to Kikori to prepare for an official visit in mid-January (1929). Soon, Musa Chilwa had my best antelope heads, beautifully polished, hanging in the verandah: at each end a handsome local mat was nailed to the wall and below one of them we placed a native bed, criss-crossed with Zebra hide thongs; on the floor, dressed skins acted as carpets. (Later on, when Maria saw our efforts she said that on their next visit she would bring out her sewing machine as she intended 'to make the house a home fit to bring a wife to, and not a game keeper's lodge'. She was as good as her word.)

The official visitor was Dr. W. B. Johnson, Head of the Nigerian Tsetse Investigation, which included sleeping sickness. He and Belt, a young American, had driven across Africa and were making for Cairo. Swynnerton, Phillips and Potts, the senior entomologist, came out with them. They slept in tents and used Rhino Hall in the daytime. The visit lasted a week. Johnson, a truly great man, later became Sir Walter Johnson, Director of Medical Services, Nigeria – and my boss, as four years later I transferred to the Nigerian Sleeping Sickness Service. On this visit he opened my eyes to the magnitude of the problem, as well as kindly sharing his Stovarsol tablets with me as I was far from cured.

Soon after the visitors left, I had to return to Kondoa and begin again the 12-day course of Emetine injections. It was a laborious procedure for the doctor as, having no hypodermic syringe of adequate capacity, he had to insert the needle in my arm, inject, leaving the needle hanging, refill the syringe, connect it to the needle and inject again, and then repeat the procedure once more. We were both delighted when the course was finished.

During this interlude, Noel and Maria kindly put me up. One night, Caroline, Maria's tame wart-hog, dug a hole under the outdoor privy. Maria fell into it and barked her shins. How I wished I had known Portuguese! One day, some natives, knowing Maria's love for all animals, brought her a pangolin, the Scaly Ant-Eater – an armadillo-like animal covered in horny, over-lapping scales. It was over three feet long and when frightened would turn itself into an armour-plated ball with a tremendous clash as its tail swept in. They are very rare, nocturnal, live in burrows and feed with their long sticky tongues on termites and ants. For the night, we lowered it into a 40 gallon drum, put a sheet of corrugated iron on top and weighted it with a large stone, but by the morning it had overturned the barrel and disappeared.

On getting back to Kikori I found that, owing to drought in the Masai steppe, the game had come in and nightly the lions walked through the village roaring ostentatiously. Victor and I decided to sit up for them on a *machan* – a platform raised 10 feet above the ground on four legs. It was sited by a water-hole.

We had no torches as Kondoa had run out of batteries, so we had to wait for a full moon. First some zebra came down to drink and splashed and kicked each other in typical zebra fashion. Then there was silence apart from the ping of the mosquitoes or an outbreak of croaking by the frogs: once a porcupine rattled his quills and lumbered away.

At about 11.30 p.m. we heard lions grunting, apparently at a distance. We looked at each other – the lions had started to move. Only a few seconds later there was an ear-splitting roar within a few yards of us – a noise like nothing on earth. Dense black clouds passed over the moon, we could not see the ends of our rifles. A deathly silence ensued, and then a slight rustle below us and deep breathing which suddenly stopped. We sat there praying for the moon to reappear. After what seemed like ages, a silver light lit up the water-hole and we peered through gaps in the floor of our platform. There was nothing there. Next moment we heard a loud snort, indicative of fear. There was a pause, a sudden stampede of hooves, two more agonised snorts and a crash, a ripping, tearing sound followed a little later by crunching – all on the far side of a small thicket. The zebra herd depleted by one, came down and drank, barked, kicked and

76

splashed, knowing that the lions were preoccupied. Two hours later, all was quiet. (I now understood the feelings of the two society debutantes, recently out from home, who had sat on the ground within a thin thorn fence listening to those same noises – augmented by the snores of their drunken white hunter.)

At this season the Kikori villagers would track down lion-kills by watching the vultures. If a lion was present, they would climb a tree and yell and yell and yell, until the poor beast's nerve was shattered and he cleared off; they would then descend and make off with the lion's meat. Once I met three would-be meat stealers, drooling at the mouth with fright: they had found an untended kill: one of them had just picked up a zebra's shin bone, when there was an outburst of growling: they shot up the nearest tree, as eight lions strolled out from an adjacent thicket and resumed a leisurely meal – with a captive audience.

Life was quite amusing in camp, despite one of my *askaris* who owned a bugle and would wake us up at dawn by sounding the 'Last Post'. At this time I wrote to my brother in Ceylon as follows:

'You may think that life in the African bush is quiet. Well, so it is if you don't keep a puppy and a young monkey who dash round and round the house for hours on end and smash all the geranium cuttings that have been so tenderly nurtured. Guess what I found under the bed last night? Wildebeest balls in their bag. Apparently one of the *askaris* was keeping them until high – as a special delicacy – and the puppy stole them and hid them under my bed.'

The puppy was a fox terrier type, but sandy and white instead of black and white. When good I called him *Mbwa* (dog), when bad *Mbweha* (jackal); I regret to say that the latter became his name. However, the *askaris* called him *Jumbe* (chief), as they said that he was by far the most important person in the camp: they also claimed that he was a member of the Masai tribe, as like them, he revelled in drinking fresh blood, which he did whenever I shot an animal.

One evening at dusk I saw two huge bull giraffe in the middle of the road, each some 18 feet high. I got to within 15 yards of them when Mbweha gave a yap and charged. The giraffe swung round and lumbered off with that extraordinary gait – both left legs forward, then both right – hotly pursued by a yapping little terrier.

What a joy it was to get back to camp after doing a fly-round in the heavy rains, soaked to the skin and very tired, to find hot tea already poured and a mango on a plate, during the eating of which Musa would remove my leggings and mud-clogged boots, whilst the

77

waterboy prepared a bath. Musa, full of news, would tell me how the monkey had killed a duckling, or eaten a geranium, and would always finish up with '*Bwana*, the children were very bad today, Mbweha barked at everyone.' What a simple life it was, no wireless, no news – just a dove that cooed . . .

> 'Some birds lay one egg
> Some birds lay three,
> But I lay – too' ta' too' – two'

Monotony was broken by a swarm of locusts, flying very low with a whirring of wings. The air seethed with them. Some settled in my vegetable garden. We rushed down, beating kerosene tins, and drove them off – unfortunately onto the headman's sugar cane, but he drove them off onto someone else's farm. The locusts then rose and flew out into the plains.

One day, after a two hour stalk, I shot a magnificent bull oryx; the horns were 32 inches long – slender and tapering. I was far out in the steppe and would never have got a shot had it not been for Saidi, my guide. The herd was upwind of us in an open, short-grassed plain. The oryx were about a quarter of a mile away and partially obscured by an intervening troop of zebra. We crawled to within thirty yards of the zebra and were lying under a bush when Saidi produced a couple of small stones and very gently tapped them. The zebra looked up, were inquisitive, came a little nearer, then grazed. Saidi tapped again, the zebra looked up. He repeated this performance until the zebra became somewhat uneasy and changed the direction of their grazing. We let them get some distance away before completing our stalk.

Soon after this incident I adopted two techniques when hunting. I always took a forked stick on which to rest my rifle, as owing to a football injury to my left arm I found that it would not take the weight for a standing shot without quivering; this was especially the case when using my heavy 404 Mauser. My second technique was ludicrous, but it often worked. I came to the conclusion that in open country, where long stalks were needed, a thought-transference took place between me and my prey, which caused it to become prematurely alarmed. I therefore gave up crawling in a straight line towards the animal. Instead, I would get closer by casually walking in an oblique direction whilst doing the 13 times table in my head, which cleared it of ill-intent. An occasional squint through the corner of the eye, and then when within range, up and bang. The nearer one gets, the more difficult is the table; by 15 times 13, I became innocuous. One only has to see a gorged lion walking through a herd of grazing zebra which don't even bother to lift their heads, to realise that there may be something in this idea of thought-transference creating alarm in the hunted.

The black-necked or 'spitting' cobra ejects its venom in a fine spray with sufficient power and accuracy to reach its enemies' eyes at a range from six to nine feet. I knew several old Afrikaners who always wore a highly polished silver button in the centre of the belt, which they claimed would deflect the aim of this cobra from the eyes to the button. The venom temporarily blinds the enemy, and according to the natives, milk is the best antidote.

One day, Musa rushed up to me with Mbweha in his arms, saying that a cobra had spat at the puppy. I took one look at his eyes and sure enough they were bloodshot and almost turning inside-out. Owing to the tsetse fly there were no cows and hence no milk at Kikori, so I rushed down to the village with Mbweha, found some women and shouted 'quick, two shillings to the woman who will give my dog a squirt of milk in each eye'. Shrieks of delight from the women, a young nursing mother was soon pushed forward and kindly obliged whilst grumbling at the waste of good milk on a dog. Poor girl, she was always having her leg pulled as being foster-mother to the *bwana's* dog, but she took it very well and greatly enjoyed her notoriety. The antidote was rapidly effective.

Whilst on the subject of snakes, the villagers used to 'fish' for pythons in the stream using a grappling hook, inserted in a lump of meat and tied to the end of a rope. They said that they killed the python with poisoned arrows, but I never saw this done.

79

CHAPTER 6

A Shooting Safari into the Masai Steppe

As I was entitled to three weeks local leave I decided to revisit Misoui in the Masai steppe, and this time to go further afield. However, I found that I would not be able to afford the cost for more than 12 days as even this brief period in uninhabited country would necessitate 25 porters, 12 of whom would be carrying flour. Since on my previous visit I had had trouble in finding water, this time I picked early May – the end of the rains.

The country I wanted to explore lay to the east and north east of Galappo and Ufiome Mountain; it is now well mapped and forms part of the Tarangire Game Reserve. I shall use the names of the hills and camping sites as given to me at the time: very few of them appear on the current maps. The following is an abridged version of the account which I wrote on the *safari*.

3/5/29

We set off at 7.00 a.m. The party included 10 Kikori villagers who wished to join me on their fishing and honey-hunting expedition, as the country to which we were going had a bad reputation for lions, rhino and elephants, and also the escarpment natives still had a sneaking fear of marauding Masai who in the past were always raiding and pillaging their villages. The morning was damp and cold and often there was thin rain, which rather depressed the porters.

Having reached the foot of Mount Ufiome, we struck out into the Masai steppe, following a track that meandered for 5 long days, until Arusha was reached. Much of the way led over undulating open country, each of the hills being named after Sultan Dodo's brothers, who were hung, one on each hill for their share in the Wa-ufiome rising against the Germans. This old track which was used as a war path by raiding Masai, as a slave route by the Arabs, and as a road by the English in the war, must have seen much misery and suffering.

80

Just before reaching 'The Camp of *Bwana* the Doctor', I shot a bull hartebeest. This camping site is named after a German Veterinary Officer who, years before the war, followed the caravan route to Arusha, mounted on a horse and accompanied by his wife. Since the war, few, if any, Europeans had used this route.

Having got into camp, I drank gin, played the gramophone and looked at Ufiome Mountain with the setting sun showing it up in strong relief, and very beautiful it all seemed after a 20 mile march. Then came the rationing, each man being given a carefully measured tin of flour. Watchfulness is needed in uninhabited country, as if the flour runs out the porters bolt and leave you stranded miles from anywhere. A favourite trick is for a porter to get his flour ration, and then come back for another; with 25 strangers it is difficult to recognise a man who has already had his portion.

For fear of lions, everyone kept close to my tent. There was a squelch, squelch as meat was guzzled, followed at about 9.30 p.m. by awful hiccups which gradually merged into shattering snores.

4/5/29

We struck camp at 6.00 a.m. and reached Mtaringire at 4.00 p.m. Shortly after leaving camp, we saw the spoor of two bull elephants, but it was about 24 hours old. Having passed Misoui, we left the open savannah and entered country that was new to me. It was horrible. The track was sunken and about a foot wide and walled in by 6 ft. grass. The path went through thorny, spiky trees that overgrew dark thickets. There was rhino and elephant spoor in profusion, but nothing of any size. All day the sky was overcast and there was a thundery breathlessness in the air. After 24 miles or so, we stopped at 'The Camp of the Wife' and had lunch. The site is so named because the German Veterinary Officer's wife slept here whilst her husband slept at my last camping place. He had promised to catch her up, but in his efforts to shoot some game he was so delayed that he had to sleep 14 miles behind her. The lions roared all night and terrified the poor wife; when her husband arrived next day she gave him a dressing down in front of all the boys. Such is the local story, told with much gusto.

Shortly before getting into camp, a herd of impala crossed the road at full speed, bounding over the long grass in the most superb manner, head and neck thrust out with horns lying flat along the back. No greyhound was ever so graceful.

At last Mtaringire camp was reached, the noisome thorn thicket left behind and in front, green parkland dropping to a winding river.

81

5/5/29

We started at dawn and followed the river northwards. I had rarely seen anything more beautiful than this strip of green parkland winding through a sombre forest of flat-topped trees. Along the banks were clumps of beautiful 40 foot high palms with slender black trunks and crowns of pale green. They seemed to tower over the squat forest on either side. The whole scene was incongruous: an English park with a winding trout stream, bordered by delicate palms with African giraffe beneath them, and all in a coal black setting. Buffalo, rhino and elephant tracks were numerous. We left the river for a while and plunged through thickets and shrubberies, and during this stretch my riffle was very much to hand. We got through without adventure after two hours of terribly slow progress, and reached the proposed camping ground. To my dismay, I found that this part of the river was dry. Fortunately, after a few minutes, we found a hole which elephants had dug with their tusks the night before in their search for water. We camped here because we did not want to go too near the home of the elephants, which was said to be at the foot of a hill some 3 miles away.

As I wrote in my tent, I sat facing the promised land. It was the most fascinating volcanic type of hill, of the weirdest shape, supporting only an odd tree on top, and with dense thicket at the foot.* Next day, I hoped to find a good bull elephant.

6/5/29

We started at dawn, to the accompaniment of a disappointed and hungry lion who roared in disgruntled fashion about a mile away. For hours we crept through the dense thicket, with a visibility of about 5 yards. A nerve-racking game when one knew that only the greatest care could save us from blundering into the cow herd when there would be hell to pay.

Everywhere were traces of yesterday's visitation. Trees torn up by the roots, branches broken, holes dug in the ground for succulent tubers, and winding paths smashed through the thickets. In one place where an elephant had wanted to get down into the river bed, it had caught hold of an over-head branch with its trunk, so as to steady itself, as it slithered down the steep bank. The branch had broken and, from the churned up sand at the bottom, it must have come down most ungracefully. All the spoor was clearly that of cows and

*Believed to have been the Tarangire hill of current maps.

82

calves, with the exception of one huge bull who had followed about 5 hours after the rest, and about 6 hours before us. We followed his spoor right round the hill. We saw where the elephants had crashed up and down the river banks in search of water, where they had slithered down and dug with their tusks in the dry sandy bottom, and where they had scrambled up the precipitous sides. Finally we found tracks leading from the hill and heading straight for their home (Mbugwe) some 40 miles to the north-west. I gave up the hunt and sat down for the first time for 8 hours.

The guide then lost his way in the dense bush, and everyone suggested a different direction. If it had not been for my father's pocket compass we would have been lost and have had to sleep out. I took the direction which everyone said was wrong and, on reaching camp, everyone said it was the way *he* had advocated. They argued about it for hours.

After supper the porters hovered around in the hopes that I would play the gramophone which to them was a mystery closely related to witchcraft. They would sit around and laugh and laugh until the end of the record. Goethe's *Faust* was their favourite. There was wild excitement when it came to the inane laugh of Mephistopheles – 'Shetani' as I described him. Time after time I put on that record.

As the elephants had left the area and as the seven feet high grass precluded the shooting of antelope, I decided to set off at dawn next day in search of fresh hunting grounds, as the carriers were clamouring for meat.

7/5/29

When we set off the sunshine was brilliant, but it was exceedingly cold until 7 a.m. The high grass was saturated with dew, and I was soaked to the skin until about 9 a.m. when the grass dried and the weather became really hot. At about 10 a.m. I saw 5 grand impala bulls, and dropped one with a heart shot. One still remained facing me in the long grass, and as we would still be short of meat I put a bullet into his neck, but missed the spinal cord. We followed him for four hours when we lost the tracks on hard ground; I felt badly about it. This diversion made us so late that I decided to spend the night at 'The Camp of the Wife'.

8/5/29

It was an awful day. We did about 30 miles. Camp was struck at 6.15 a.m. and by 2 p.m. when we arrived at our proposed camping place we found that there was no water. We had to walk until the late afternoon when at last, as a result of following game paths, we struck a puddle of cocoa-like fluid. Never was water more pleasant – it made excellent tea though 'of the earth, earthy'. We had had to go miles out of our way to find it and were now once more among the foot-hills of Ufiome Mountain on *Mpimbe hill.

Although almost dusk the view from our camp was superb. We were perched on the lip of an extinct crater to the east of the mountain which appeared black in its mantle of rain forest. Turning round, we overlooked the Masai steppe – huge grassy plains interspersed with dark islands of thorn scrub. The sun set; it was cold. Suddenly Kilimanjaro appeared – some 130 miles away – lit by the sun we had already lost, which turned pink its snow-capped peak – nearly 20,000 feet in height. The vision lasted but a minute: how apt its name – The Mountain of God. Around our crater were little grass-topped pimples, relics of Ufiome's volcanic period.

Earlier, I had shot a hartebeest, so we all dined well.

9/5/29

Today we walked north-westwards for some eight miles across the foot-hills of the mountain. Our destination was Mamira, a hamlet with about 15 inhabitants in the spirit-ridden area referred to earlier, p. 57. We followed a path that was over a foot deep and a foot wide – a sort of drain. The grass was about 8 feet high and completely concealed the track, so that we walked by 'feel'.

On arriving at Mamira, we found that it was deserted. Every hut was empty. The entire population had fled when they saw us coming. Later, my *askaris* produced one man. I talked to him. At first to all my queries about elephant and rhino, he replied 'there are none, you had better go elsewhere.' His attitude was distinctly unfriendly. I then told him that the man who showed me an elephant would get 10/-, a rhino 5/- and a buffalo 2/-; after this he bacame slightly more helpful, and said that he would look for a guide.

I was probably the first European to camp here and they certainly did not want me. They were reputed to have very powerful arrow poisons and I began to wonder whether they were engaged in killing rhino and elephant on a large scale and selling the horns and tusks to the Arabs, in which case they would not want me to come across the skeletons. They were certainly very sullen and unwelcoming.

*Probably Tsarmai hill on current maps.

There was quite a scare last night. I heard a tremendous crashing and the noise of snapping branches coming down the hill behind my tent. When the sounds became louder I thought I had better get up and see what was going on. The camp was deathly quiet, suggesting that everyone was lying with his head under his blanket, hoping that the wild beast would take his friend and not himself. I collected two *askaris* and armed with my heavy rifle and torch we went to investigate. The *askaris* swore it was either one elephant, two rhinos or a herd of buffalo. I had no intention of following the beasts, but wanted to know whether a mad ill-tempered rhino was considering charging through the camp.

When the noise was close, I switched on the torch. There was a wild stampede and a herd of zebra dashed off into the night. The hill was covered with thicket and loose stones, and in their descent the zebra had made enough noise for at least a couple of rhino.

The villagers seemed to have got over their fear, and two guides turned up in the morning, rather the worse for a village 'blind' the night before. We found much rhino and buffalo spoor of yesterday, but nothing fresh.

We were crawling up a precipitous hill, clothed in dense thicket and twining creepers. For hours we had been crawling and squirming through this miserable undergrowth, never saying a word. We were strung to a high pitch; I never ceased to caress the safety catch of my rifle. Suddenly a stealthy rustling movement began above us. The sound grew louder. A herd of animals was descending on us. A suppressed series of deep grunts confirmed our suspicions – buffalo. A precipitous hill, clad in impenetrable undergrowth, was a most unsuitable place to meet a descending herd of buffalo. On getting our wind, they would probably stampede, and if so, would certainly come on down, as the steepness would prevent their making a rapid retreat. We got behind a fairly thick tree, and I waited with safety catch off and barrel resting on a nice horizontal branch.

I decided in my mind just where I would aim, through the tiny gap between the base of the two horns. The noise seemed only a few feet away. The suspense was great. Would those bushes never part and reveal a fine pair of horns, vindictive blood-shot eyes and a slobbering muzzle?

At last the bushes did part, and I stood face to face with a huge baboon! There was a tremendous grunt followed by blood-curdling

85

screeches, cries and chatterings. The females and their children squeaked and there was a stampede uphill. Two ludicrous scares within twelve hours. It was high time for lunch.

In the afternoon, we followed the guides to a spot where they swore that a large bull rhino always slept. They warned me that the grass was rather long. I was desperate to get something, as so far I had had no luck with big game at all.

The guides had not exaggerated – the grass was at times 10 feet high. I told them that it was not good enough, but they said that the rhino lie-up was in a place with no grass, only a few yards ahead. They pointed to some unpleasant tangling creepers and said 'in there *bwana.*' Between us was a deep gully. The guides said to go a few yards to the right and cross by the natural bridge. Realising that the wind would blow our scent down into the thicket if we did this, I led the way down the steep sides of the gully and had only just got up the other bank to the edge of the thicket, when I heard three colossal snorts; there was a crash and I got a glimpse of a rhino charging by. He dashed straight for the bridge, where it turned out that one lazy hanger-on had decided to cross. The rhino had got his wind and charged him. I couldn't see to shoot. Apparently the native jumped for his life over the side of the natural bridge and rolled down into the gully, whilst the ill-tempered rhino rushed across. The beast had a miserable horn and so fortunately never came back. I thought we had had enough rhino and thicket for one day, so we returned to camp.

11/5/29

These local Wa-ufiome natives were certainly very strange. Yesterday I mentioned to Kompania that my bad luck was extraordinary. For 8 days I had tramped the bush, crawled through awful thickets, and yet had shot no big game, nor even seen any apart from the rhino. Quite often we had found fairly fresh dung and spoor, but never had we succeeded in tracking the beast down; the spoor had always disappeared suddenly. Saidi, my Kikori guide, who had worked for me for 18 months and in whom I had great faith, heard my remarks to Kompania. He turned round quite fiercely and said: 'It is your own fault *bwana*, you don't understand how to hunt in the Ufiome country.' I asked him what he meant. He replied that I had come into the district and instead of asking for a local guide to show me where the zebra and hartebeeste lived, I had openly announced that I wanted to shoot rhino, elephant and buffalo. Now the Warangi, to which tribe Saidi belonged, know that the Wa-ufiome are evil people who practice the worst forms of witchcraft. He went on to tell me stories which I had heard before. – The Wa-ufiome have a powerful medicine with which they can capture elephant, rhino and buffalo in a few minutes. Each big witch-doctor owns his own herd and marks it, native fashion, by an arrangement of slits in the ears; the bull and most of the cows are thus marked. Further, these magicians can call up such an animal and ride upon it to the house of an enemy and then kill him. Again the females of these beasts can be milked and a terrible poison concocted using milk as one of the ingredients. When a chief dies his spirit enters a previously selected animal. Saidi went on to say that the reason that we could not see any big game was because on the day of our arrival the Wa-ufiome had sent word all over the district that a white man had come to shoot elephant, rhino and buffalo. That night all the owners of these wild beasts had made a medicine which would prevent any of my guides seeing one even if he was close to it. However, a European is supposed to be immune to the charm.

On hearing this, I said that white men had brains and didn't believe in the gossip of old women. Saidi remarked, white men are different, but black men know this to be true.

Last night was noisy. A hyaena whooped in the village and a lordly lion roared from 1 a.m. until near dawn with cheeky Mbweha barking back. In the morning there was lion spoor all around the camp.

We set off at dawn but failed to see the lion. At about 8 a.m. we were walking through orchard bush country – orchard sized trees with grass growing between them. My thoughts were elsewhere and I was relying on Saidi and Kompania to spot any game, as both were remarkably keen-sighted. I chanced to look to my left and saw two huge black beasts grazing only a hundred yards away. I whispered 'rhino' and pointed, and only then did they see the animals. How had they failed to see them? I suspected that their belief in the vision-inhibiting medicine of the witch-doctors was the cause.

I looked through my field glasses to see if the horns were any size – they did not appear to have any horns. Saidi then said rather dreamily 'possibly they are buffalo.' He was right. Never having seen a buffalo before I had not realised what a massive beast it is. I fired, and as I later found out had hit him in the heart region. He turned and came lumbering down on me. He paused for a second, either to see if the female was following him or because he was feeling sick. I fired again and smashed the shoulder. He fell. The female looked very threatening, but I did not want to shoot her. After some time she

87

moved away, so I approached the male, but kept a close eye on her. The bull was dead. A huge animal and very old. His horns would have been magnificent had they not been so chipped and worn. The outside span was 45 inches, and later it took four men to carry home the head slung on two poles.

I was examining the head when Saidi turned reproachfully to me and said 'Now *bwana* you see that I don't tell lies.' He pointed to the ears. The right-hand one had five V-shaped slits comparable to those made in the ear of an ox, but the left-hand ear was shredded by age. (I had the head mounted in England; it is now in the Haslemere Museum).

Having shot the bull, I left a man to keep the vultures off the meat and went in search of a rhino. Having returned empty-handed, the watcher told me that he had been driven up a tree by the female buffalo, who had returned. She left, having found that the bull was dead.

It was only at this stage that I learnt that the two guides produced by the Mamira headman were not Wa-ufiome, but Warangi; his own people had refused to help me, lest I shot one of their branded beasts. When we got back to camp the villagers were furious with the two guides and threatened to beat them. The moment my buffalo head was brought into the camp, the hamlet headman came and studied the ears. I asked him how his people captured buffalo. He looked stupid and said 'Possibly, perhaps, possibly,' which did not get me any further. I learnt next day that in killing the buffalo, I had destroyed the resting place of a dead chief. It was all very weird and very strange, but buffalo often do have torn ears.

12/5/29

As my camp was two hours walk from what was reputed to be the best rhino country, I decided to leave the main body of porters at Mamira, take a groundsheet and a few essentials and walk to the rhino habitat. Within half an hour of arrival, a nice grass hut had been made for me on top of a hill. That evening a gale got up and threatened to blow over my flimsy shelter, but it failed to do so.

13/5/29

I started hunting at dawn, but never saw a rhino because a crowd of Wa-ufiome, on the pretext of cutting grass, made as much noise as they possibly could in the vicinity of the rhino thicket.

We were in a huge dense forest, with creepers and thorns, when we disturbed a large herd of buffalo. We never saw them but heard them crashing off, some uphill and some down. Fortunately they were content to remain separated until we had passed between them, but then half way through we found tracks of a rhino and calf going in the same direction as we were: however, we did not meet them.

14/5/29

We returned to Kikori after an enjoyable, but very poor shooting trip, due to constant bad luck and alleged bewitchment. It was good to breathe the spirit-free air of Kikori, apart from the evening fumes of gin or whisky.

I still have a memento of this trip – my moustache. For twelve days I had not shaved. When I looked in the mirror I saw a revolting, seaweed-coloured, vestigeal beard, from which I was soon, but painfully, parted; my cut-throat razors had not been set by a barber for one and a half years. (Incidentally, Musa always cut my hair; the end result resembled a relief model of the rift valleys).

CHAPTER 7

Visitors, and Other Happenings

During the next four months we had a spate of visitors from abroad, attracted by the new road which had opened up this little known area. The first was Bailey Willis, Research Associate, Carnegie Institution of Washington. He was a seismologist who intended to study the Rift Valleys in connection with earthquakes. The rifts are believed to represent fractures in the earth's crust. He and his colleagues were to examine 15 hundred miles of these long, narrow depressions.

Bailey Willis, who had a magnificent white beard, was accompanied by Dr. Teale, Director of Tanganyika's Geological Survey. They came up to Rhino Hall for drinks, and we had a most amusing evening. In his subsequent book, 'Living Africa', there is a reference to myself. – 'He entertained us with his enthusiasm for hunting and his stories of game. An enraged python had chased him a mile – and it was not a bootleg story either.' (It took me a long time to live down that remark.) They left next morning, visited Galappo Mission and the small cemetery with George Rex's grave, saw Ufiome Mountain, and thence to Babati.

In the Preface I stated that I was going to keep off the subject of tsetse research and deal only with the background to the work; however, there is one series of experiments which may be of general interest.

I wanted to gain an insight into the movements of individual tsetse, and so caught wild flies, marked them, let them go and hoped that some of them would be recaptured. Fortunately my tropical outfitter in England had provided me, among other useless items, with two tins of Blanco – a white cake, provided with a sponge, which was intended for whitening tennis shoes. Having ground up a cake, I produced different coloured powders, using mapping inks as dyes. The powders were wetted to make a cream before application to the back of the tsetse fly. The insects were restrained for a minute or two in U-shaped pieces of plasticine, until the paint had dried; they were then released by opening the 'U', when they would dart away.

Laboratory experiments had shown that this treatment did not adversely affect the insects.

Two very small pools, in a thinly wooded vlei or '*mbuga*', were kept under observation for three years. In the early dry season these pools retained their water for some weeks after all other pools had dried up; in the mornings and evenings, hartebeest, wart-hog, zebra and giraffe would arrive and await their turn to drink; there was a large concentration of tsetse flies around the water-holes. Then the pools dried up and the game left, but the fly concentration persisted for about three weeks and dispersed in the fourth.

Several days after the pools became dry, the flies were marked and released. Surprisingly, it was found that the fly concentration consisted of ever-changing individuals – rather like an airport hotel. The flies were very hungry, and attacked furiously. It was postulated that a stream of hungry flies followed the game paths, which converged from all directions on the pools, and that having waited some time for game to arrive, the individuals flew off; further, that with disuse the grass fell over the paths, the scent disappeared, the paths ceased to attract flies, and the concentration disappeared. (It was assumed that only half the hunting flies, attracted by a path, followed it in the right direction.)

In the following wet season, when game was widely dispersed and there was no fly concentration at the pools, six imitation game paths, totalling four miles in length, were cut to converge on one of the water-holes, around which a small clearing was made; both paths and clearing were hoed. It so happened that one and a half inches of rain fell on the night after the work was completed, which should have removed any scent left by the labourers. Within a few days a tsetse concentration formed around the pool on which the paths converged, but not around the second pool; again the population consisted of ever-changing individuals.

The results indicated that one method by which tsetse locate their hosts is by following paths.

Whilst I was engaged in this investigation Swedi Abdallah was seconded to me. He was Swynnerton's Head Fly-boy – thick-set and with an awful stammer, but most able as I soon learnt. On the day after his arrival we were both sitting round a camp-table under a tree,

91

and applying colour marks to the flies. 'Swedi,' I said facetiously, 'if my red male mates with your blue female, what colour will the offspring be?' Looking quizzically at me, he replied, 'Who knows, *bwana*? it is the affair of God, but why does a black cow give white milk?'

Soon after I got back to Kikori, I was visited by Dr. L. S. B. Leakey, the famous Kenyan archaeologist and excavator of the Olduvai Gorge; he was accompanied by his wife and two friends. They were on their way to a meeting of the British Association for the Advancement of Science which was to be held in Johannesburg. They had deviated from their route to see the rock paintings which I had found. The visit was described in Leakey's book 'White African'.

My chief memories of this enjoyable interlude was Mrs. Leakey cutting my loaf of bread with a Stone Age spear-head made from black obsidian – a volcanic glass – which was part of a collection of weapons to be exhibited at the meeting. Presumably obsidian is easier to shape than flint, as the weapons were most beautifully made. My second memory was of Leakey's story of how he had shot a large python which seemed to be dead: he had put it in the boot of his car, but on getting home the snake was found to be very much alive.

The opening of the road and the start of a research station by Phillips, in the hills on the far side of Kikori village, led to uncontrolled hunting, which threatened my studies on game movement in relation to tsetse. I succeeded in getting the whole of the Kikori area turned into a game reserve, and left it to Victor, with his car, to get meat for the staff from further afield. In consequence, my Ufiome buffalo was the last beast I shot before going on leave six months later. It was a busy period as I was writing up my research results for a Ph.D. thesis.

On an earlier occasion I went out with Kompania to get meat for the staff. I shot a hartebeest not far from the camp, and then sat down in the shade of a very small thicket and smoked a cigarette. I then left Kompania to keep the vultures off the kill whilst I walked back to send out porters for the meat. I had only left about five minutes before a lion strolled out from the tiny thicket at the edge of which I had had my cigarette. The lion sat down and faced Kompania, licking his chops and regarding the antelope. They sat facing each other, Kompania holding his little one and six-penny knife and *refusing* to leave the meat. Finally the porters appeared and the lion bolted. Such was Kompania's story, and I believe it; he was utterly fearless – embarrassingly so, as on the occasion already described when he pursued the rhino and calf. I suspect that the lion had been watching the hartebeest when we first saw it, had hidden in the thicket on our approach, and had only emerged after I had left it and conversation ceased.

On another occasion I shot a large zebra late in the afternoon when I was far out near the edge of the steppe. I left two or three men to guard the carcass overnight, having given them matches so that they could light a fire, and set off alone for Kikori. I had a rifle and could not possibly get lost as I only had to walk in a straight line towards a land-mark on the face of the escarpment. after about an hour, I suddenly realised that I was often looking over my shoulder: the intervals lessened as time went on: I began to feel that something was following me and was glad when I neared the village. Only then did I realise why the fly-boys were unwilling to go to bush singly, and always insisted on taking a companion to whom they could talk. Next morning I returned to the kill with enough porters to bring in the meat. On arrival, I was amazed to see how little meat was left: the men had eaten and vomited, eaten and vomited, throughout the night. Only then did I fully realise what meat-hunger means to those living in tsetse belts where cattle cannot be kept.

Mbweha, my dog, always accompanied me to bush. One morning he seemed to be rather off colour, so I left him behind, shut up in my tiny laboratory cum office. On returning from the fly-round I opened the door to see how he was. He came straight for me foaming at the mouth, with protruding greenish eyes. He seized me by my leather-gaitered leg. I kicked him off with the other boot, slammed the door, got my gun, re-entered, and shot him dead. He obviously had rabies.

Thereafter, I hated going into that room where he had always lain by my feet as I worked, but fortunately a larger, detached laboratory was soon to be built. Musa was most upset, and the whole camp mourned that little dog, as did Tumbo – the monkey.

No one had been bitten by the dog, so I did not worry about hydrophobia. Later, in Nigeria, I discovered that even if one had only a scratch on the hand which might have been licked by the dog, the doctors insisted on a full course of treatment consisting of large injections in the stomach muscles administered for some weeks.

For the next forty years I followed Kipling's advice:

> 'Brothers and Sisters, I bid you beware
> Of giving your heart to a dog to tear'

The Africa I knew was no place for dogs. In the towns, as soon as a case of rabies occured, an order was issued that all dogs must be kept on leads for six months, but there was always another case before the sixth month expired.

As I write this book, deliberately based on my letters home, I am appalled at my ignorance as a young man. I must have been more ignorant than many, due to the fact that I lived in bush and belonged to a newly created department in which we were all young and ignorant. Most young men had the advantage of overhearing conversations by their elders in the club and at drinks parties, and soon picked up knowledge on hunting, diseases, etc. It took much longer if one had to learn by bitter experience – and depend on hindsight rather than foresight.

Soon after Mbweha's death I had another attack of amoebic dysentery. Victor kindly drove me into Kondoa, only to find that our doctor was ill and had gone to Dar es Salaam for treatment. In consequence, I had to go to Dodoma where, for the first time, the disease was diagnosed microscopically. Despite the fact that a new hospital had just been built for Africans, there was not a single bed for Europeans, so I had to stay in the 'hotel' – a dreadful place with filthy sanitary arrangements. Dodoma was arid and rocky – a town in a desert: the only 'wet' place was the hotel bar.

I had two injections a day, one in each arm. By the time the course was finished, I wished I had been born a millipede.

One evening, after dark, a lorry drove up to my house. The Runton brothers, who were white hunters, were escorting a Dr. Carnochan on his travels, and wanted to camp at my site.

Carnochan was a member of a Smithsonian-Chrysler Expedition that was engaged in collecting specimens for the National Zoological Park in Washington. Snakes were his speciality. He said that he had been made a member of the Wanyamwesi Snake People's guild and had learnt about their secret medicines, including a preparation which confers immunity to snake bite. He had made a collection of native herbal drugs for analyses in the United States.

Carnochan was a short man, with a little French beard and

protuding, brilliant blue eyes that never seemed to focus on one's face, but on some spot at the back of one's skull. He pulled down his shirt collar to reveal a few straight parallel lines, like the graduations on a scale – the marks of some secret society. He stopped my servant Musa and asked him what tribe he belonged to. Musa replied 'the Wangoni'. 'I don't want to know that, I want to know where your ancestors sacrificed to their guardian spirits,' said Carnochan, pulling down his collar. An exchange of words followed, in a dialect unknown to me, and Musa paled, looked exceedingly miserable, and stood first on one leg and then on the other.

The party left at dawn next day. when I asked Musa what had upset him the previous evening, he replied '*bwana*, that is very bad man; he knows too much.'

Some years later a popular book entitled 'The Empire of The Snakes' was written by Carnochan and Adamson.

Noel and Maria spent a fortnight with me at Rhino Hall, as he had work to do at the other camp. They brought out Michael, who was now one year old, and a cheetah cub who made little chirping noises. The two were tremendous friends. The cheetah would race up to Michael, who was sitting on the floor, leap over his head, turn in his own length and stop, facing the baby, who always 'guggled' with delight. The game went on for hours. Musa strongly disapproved: he was convinced that the cheetah was a leopard and had not forgotten our Kandaga days. The visit was much appreciated as it took my mind off a gnawing toothache.

Victor had made friends at Babati, where there were a few settlers, and took me over to spend a night with one of them. The barely motorable track went from Galappo across the relatively civilised western flanks of Ufiome Mountain.

We were bumping along, Victor with his eyes glued to the track, when I shouted 'Stop!' and seized my rifle. To the right was a scantily clad white girl with some almost naked Wa-ufiome dancing round her and apparently prodding her with their spears. Then a shout from the left, 'Get out of the light, you bloody fools.' There was a UFA cine-team furiously filming the maiden in distress. We got out of the 'light' and got out of the car. They were all very friendly and we were introduced to a Count Johnson Noad – the producer I think – and to young Amery, known as Puss in Boots.

They gave us a drink and said that we must wait and see the damsel's rescue. Soon a light aeroplane appeared, and made a very bumpy landing: the pilot jumped out, ran towards the girl blazing away with a revolver: the spearmen collapsed – with laughter I suspect: the pilot slung the maiden over his shoulder, rushed back to his plane and took off. (The pilot was Udet, the famous German war ace.) How true is the saying '*Ex Africa Semper Aliquid Novi*'.

We continued our journey to our host's farm. He had a foreign guest staying with him, with a name like Zech, and was expecting several friends to come in for dinner.

Plans had been made for Zech to shoot a lion. Shortly before dusk we escorted him to a *machan* that had been built for him in a tree. Having tethered the goat we left him, and went back to greet the guests and settle down to drinking. About an hour and a half later we heard a single shot.

Before long, Zech appeared, looking very pleased with himself.

'I 'av shot zee lion.'

'Was it a lion or a lioness?'

'I do not know zat; I no see ze lion.'

'How do you know that you did not miss?'

(Haughtily) 'Be 'coz I am a Swisse Militaire. We no miss.'

Full of drink, seizing the odd rifle, we bundled down in our cars to the site. He had not missed – but it was a hyaena.

In 1926, I had attended the Marine Biology Course at Plymouth with Ommanney, a fellow student, who later became an expert on whales. We lodged in a very small hotel and found ourselves seated at a table in the dining room with Professor Julian Huxley, grandson of the famous naturalist T. H. Huxley, and himself a distinguished biologist. The other small tables were mainly occupied by worthy widows, who shook with mirth as Julian regaled Ommanney and myself with his stories: he was a great raconteur and I am sure raised his voice, so that all would be entertained.

Our next meeting was on September 16th 1929, and is best described by Huxley himself in his book 'African View'. –

'. . . It is a weary way over a bad road. At last, after ten hours driving at an average of 11 m.p.h., we see a light – Kikori, the entomological station.

'I knock at a door. A voice says come in. There on a bed, pale and big-eyed from dysentery, is a young man. He is Nash, one of the research workers. Three years ago I shared lodgings with him when we were both working at the Plymouth Laboratory: and I had no idea he was here. Tents are ready for us; and we are ready for bed.'

Professor Huxley (later Sir Julian) had come out to East Africa at the request of the Colonial Office Advisory Committee on Native

96

On safari *with Swynnerton : his experiment with intestines filled with poisoned blood.*

Saidi Abdullah and Kompania Goynando standing behind a wooden sugar-cane crusher.

Zebra Hall, Kandaga.

From outside. Note mosquito netting over doorless doorway and flat mud roof.

From inside. Note camp-bed with mosquito-net, groundsheets on mud floor, and roof-supporting trees on which clothes and haversack were hung from nails.

Rhino Hall I, Kikori. A living room cum laboratory 12 ft long by 6 ft wide. (A tent served as bedroom and bathroom.)

View from Kikori of escarpment and Kisesse pass with miombo *woodland (*Berlinia*) in the foreground.*

A Wa-ufiome guide. N.B. evergreen mountain rainforest.

Mbweha (dog) and Tumbo (monkey) feeding together.

Rhino Hall II, Kikori.

From outside.

Right on entering verandah. (After marriage.)

Picnic on top of the escarpment, Moggridge, Wendy, Burtt, Noel Vicars-Harris.

Musa Chilwa as 'King of the Dance'.

Dance by the Wa-ufiome at Bereku.

Maidens wearing brass coils.

The participants.

A bush fire at night.

Our last safari – Kikori to Dar es Salaam. Musa Chilwa, Katherine, Wendy, 'Fat Fatima', Arser.

Education. He had come to Kikori to attend a meeting with Mr. Swynnerton, Dr. G. A. K. Marshall (later Sir Guy), Director of the Imperial Bureau of Entomology, and Dr. W. A. Lamborn who was in charge of the Tsetse Investigation in Nyasaland (now Malawi). Huxley's comment in 'Africa View' on these three people and on the meeting are worth recording:

'It is a remarkable and interesting fact that these three men, all distinguished scientists and all influential in the practical field, one of them occupying the most responsible post open to an entomologist in the Empire and perhaps in the world, began as amateurs, and have none of them had any special training. I spent the morning with them and Phillips, head of the Kikori Station, listening to and discussing a long scientific report which Nash gets up from bed to read. It is curious to foregather thus, a band of biologists, in the heart of Africa, 150 miles from any railway.'

(My next meeting with Julian Huxley in Africa was in Northern Nigeria in 1944, when he visited me at the Anchau Sleeping Sickness Settlement. I said to him 'I suppose the B.B.C. let you know beforehand the questions that are going to be asked in a 'Brains Trust Programme.' He replied 'What a dreadful idea. How could one get over any impression of spontaneity?'.)

After the visitors left, I had to wait a few days for the doctor to return to Kondoa. When I saw him, he packed me off to Dar es Salaam hospital for dysentry treatment, which necessitated cancelling my passage home.

Having completed the treatment, I had to return to Kikori. I got to the station early and was waiting for the train to pull out of Dar es Salaam, when a huge young man jumped into the coach, shouted for a bottle of beer and introduced himself as Cecil Stiebel of the Administration. He was a great extrovert and very amusing. I said to him –

'Where are you going to?'

'To the Secretariat.'

'Where?'

'Dar es Salaam.'

'But you are already there.'

'Yes! Yes! but I always nip out for a beer when there is a train going up country – a brief escape to a place where nobody will find me.'

We were to become very good friends, and more of him later.

I got back to Kikori, and the last thing I did before leaving was to

97

give Victor my 37 hens and 3 cocks on the strict understanding that there must be 40 birds when I got back from leave. There were – but 37 of them were cocks.

I sailed on November 10th on S.S. 'Llanstephan Castle' which went round by the Cape, and took over five weeks to reach England.

INTERIM

CHAPTER 8

Between Tours

How fantastic it was to be on board a ship and civilised again. The food was marvellous. Beef, mutton and pork, not lean antelope whose fat, if any, sticks to the roof of the mouth; not mince, but slices of succulent meat and cold ham; fish, which one had not tasted for over two years – apart from the occasional tin of sardines: fried bacon and eggs, huge eggs: monstrous, tender chickens, not scraggy birds that barely served a meal: pork sausages, cheeses of all types, and puddings of varieties whose very existence one had forgotten.

Dinner was always eaten wearing a dinner jacket, not khaki shorts and mosquito boots. Electric lights were everywhere. No oil-filled, insect-attracting hurricane-lamps which led to the enrichment of the soup.

Plumbing was ubiquitous. One could turn on great gushing taps of salt water until the bath was overflowing, and lie full length and soak; no sitting in a hip-bath, sited away from the lamp, lest the colour of the water disturbed thoughts of cleanliness. There were water-closets – not pit latrines and flies.

How good it was to be able to pick and choose one's friends from among scores of people – not from ones and twos: to listen to the chatter in the bar, and to watch people's expressions.

How exciting after Cape Town to see gorgeous girls galore, but I had only one in mind – awaiting me at my journey's end.

How restful at the end of the day to lie on one's berth and read with a light above one's head, and no mosquito net.

One had been starved of so much for so long.

What fun it was to have nothing to do but indulge one's whims. Perhaps a game of deck-tennis with a hand-thrown, rope quoit whistling over the net, or a leisurely game of quoits thrown at a chalk-circle on the deck. If hot, a plunge in the canvas swimming bath on the after-hatch, followed by a fresh water shower, and then maybe drinks with friends sitting in deck-chairs watching the sea; or maybe at noon going into the Smoking Room to see who had won the sweepstake for the ship's run over the last twenty four hours. After lunch one could sleep, rest or play bridge. After dinner there was always some entertainment: dancing to the music of the ship's band, and once on the voyage the fancy dress ball, preparation for which kept the passengers busy for days. (Having an old pair of trousers, with a tear in the bottom, I went as 'Behind with the Rent'.)

Before air mail was introduced there was little point in writing letters on the homeward voyage – they would arrive but little sooner than one did oneself. There was no Tannoy system with loudspeakers asking if Mr. X would kindly report to the Purser's office. Meals were announced by the beating of a gong. The cabin stewards were always most helpful. It was a thoroughly civilised life.

I have distant memories of the ports of call on this my only homeward voyage by the Cape. In those days Beira consisted of a single street, with big buildings interspersed by fields in which donkeys grazed. A tram drive in Durban suggested that the inhabitants all spoke with a Scot's accent. At Cape Town, Table Mountain was superb with its spreading of the table cloth act, as white clouds rolled over its flat top. Having first had a drink, two of us bravely scaled the mountain by the just-opened cable-car system; we were lucky, as next day the system broke down leaving passengers stranded in mid-air. One evening the ship paid a brief visit to Santa Cruz, Tenerife: I went ashore with a doctor who bought lots of cheap cigars: later he tried one, and nearly passed out: next morning, when his cabin steward brought his early morning cup of tea, the doctor gave him a present of the cigars, the steward thanked him profusely, and then threw them out of the porthole into the sea.

We disembarked at Plymouth, a week before Christmas day.

It was very good to be reunited with my mother and Wendy, but it was not until I walked into my father's study, that his death really sank in. At the time, I had done all the right things – cables, letters etc. – mechanically, but I seemed to be so cut off and distant that I lived in a dream world of my own.

Soon after Christmas, I entered the Hospital for Tropical Diseases in London to undergo their course for the treatment of amoebic dysentery. It was drastic but efficacious: all day we had to try and retain our Yatren enemas, and just before 'lights out' we were given our capsules of Emetine – an emetic – and a sleeping draught to try and prevent us from being sick. By 11 p.m. most of the dysentery cases were very much awake; thereafter, all was peace. The course lasted for several weeks.

It was in this ward, that I saw my first case of sleeping sickness – a young missionary. He was constantly asking one of us to put on his favourite gramophone record: writing from memory, on one side was 'The Song of a Prune' – 'No matter how young a prune may be, it's always full of wrinkles' – and on the other side 'Down in de cane-brake, close by de mill, there lived a coloured girl – name of Nancy Dill.' How tired we got of those two songs.

Soon after leaving hospital, I spent a few nights with Noel at his home in Rugby. He was to be best man at our wedding, and was most kindly typing my thesis. We were married in late February, and achieved our Golden Wedding – not so long ago.

After our honeymoon we had some wedding present cheques to spend. I was to find that I still had an occasional fit of self-consciousness – a defect that stemmed from living alone. We walked into His Master's Voice shop in Oxford Street, and I said to a girl behind a counter 'I er, er, er understand that you keep gramophones.' (Complete collapse of Wendy: girl struggles to restrain laughter.)

In early May, 1930, we sailed for Dar es Salaam on the S.S. Malda of the British India Steam Navigation Line – to give it its full title. It was very hot in the Red Sea. Stokers, from the unbearably hot engine rooms, would come up on deck for a breather, would pull up mugs of sea water on lines, and gulp the contents. This practice must have been going on for years before the introduction of salt tablets to overcome heat exhaution. At Port Sudan, we were shown the only tree by a proud resident, and bathed in a swimming bath with water so hot that we sweated in it. (The sea was too shark-infested for bathing.)

SECOND TOUR

CHAPTER 9

A Small Wife in a Small Station

After a 5-week voyage we reached Dar es Salaam where I picked up a model A, Ford car. I had ordered it before going on leave and had given the firm my design for building the wooden box body. In front, the bench seat had a removable wooden back softened by two mattress-like cushions the width of the car, with a third such cushion to sit on. For sleeping purposes, the loads and wooden back were first removed, the two back cushions were laid longitudinally side by side and pillows placed on the driving seat. Roll-up canvas curtains around the box-body could be let down for privacy or to keep out the rain. On each side of the car there was a wooden box, secured to the outside of the body, which held a gallon tin of water – for the radiator, and not ourselves. The car was always referred to as 'Fat Fatima'. I bought her on a government loan, repayable over 2 years.

The car was put onto the train. We got off at Dodoma and ran into Cecil Stiebel, who was as exuberant as ever. He cadged a lift with us to Kondoa, and regaled us with stories. Apparently, the last time he got a lift to Kondoa was on an Afrikaner's lorry; he had to sit on top of the loads at the back, which he shared with the corpse of his host's father, and other mourners.

The heading for this chapter is based on two facts. Firstly, Wendy was only 4 feet 10 inches tall when I married her but, as a result of good management, put on another half inch; secondly, on reaching Kikori I found that in seven months it had become a small station with a very social life. Dr. Phillips had concentrated the research workers near Kikori, where they were living in small prefabricated houses built two miles south of my camp in *miombo* woodland beyond the mouth of the Kikori valley.

On arrival we found that by mistake the doors and windows of Rhino Hall had been painted blue, which clashed with the curtain

106

materials which Wendy had brought out. She dug her toes in, so Noel very kindly put us up for a week whilst a top coat of green was applied; he was living alone, as Maria was following him out.

We got into our house on a Saturday. Next morning some villagers arrived and asked me to shoot a lion that had fallen into a game-pit (a deep, tapering trench, some seven feet long, dug along the line of a footpath, with a thin camouflaged cover). Apparently, the owner had heard indignant noises coming from his game-pit which was near his isolated homestead. Believing that it was the hoped-for wart-hog, he went with a blazing brand to burn off the hair of the luckless pig – a disgusting custom. On stooping down to peer into the pit, he saw two clawed paws slash up at him from beneath, followed by a dreadful roar. He hurled his spear into the pit, and tore back to his house where he lay terrified until dawn.

When the deputation arrived, Noel was adjusting the chimes of a clock which he and Maria had given us as a wedding present. He decided to come along and see what happened. After a long walk, the owner complained of a bad foot and said he must rest, from which I gathered that the lion was near; however, another follower said that he would show us the way. When we got near, I went ahead followed by Noel, but Kompania interposed himself between us. The very tall grass almost met over the path, so I approached cautiously, having no wish to join the lion. I expected to see the beast fuming in the bottom of the pit, and clawing up the sides; instead I suddenly found myself looking at an irate lioness lying on her back within two feet of the surface. I gave her three rounds rapid – one in the head and two in the chest, whilst Kompania aimed at her with my shot gun which he did not know how to operate. Thus was my new rifle, a wedding present from my brother, most ingloriously blooded on the poor lioness. She must have gone head over heels when her front feet went through the flimsy cover and become wedged on her back in the tapering pit.

I then sent a note to Wendy to bring the car along to the nearest point. We pulled the lioness out by attaching ropes to her legs and got her carried up to the road, where Wendy met us and drove the lioness up to the camp for skinning – a new 'carpet' for her new home.

My old rifle was a collapsible Manlicher .375″: by removing a pin, the stock could be separated from the barrel: owing to the play between wood and metal, it kicked like a mule and my sparsely-covered shoulder became black and blue after firing only a few

107

rounds. My new rifle, built in one piece, was a 9.3 mm Mauser with a rubber-padded butt; there was no kick and my marksmanship improved enormously; further, its striking force was much greater.

A few days after my arrival, Victor Findlay went on leave, so for the next few months I had all the fly-rounds to do as well as my basic research; but I was greatly cheered to hear that my first tour's work had been published and that I had got my Ph.D.

For the next few months we were to have a very social life, widened in scope by having a car. The building of new roads and the improvement of the old, had recently opened up the area. We could now get to Babati, 18 miles away, where there was a small settler community; a pub had been opened there, The Fig Tree, and it had a store where we could buy fresh cheese and bacon imported from Kenya. In Kondoa, Grant was now the District Officer and Robinson the A.D.O; both were most helpful administratively, and delightful socially.

To my great relief both Musa and Hamisi, the cook, got on splendidly with Wendy. (So often the servants from a bachelor's home resented the intrusion of a wife.) Within ten days of getting into our house we had the whole station to dinner – seven of us. The menu consisted of soup, harte-beest liver with cauliflower-au-gratin, tomatoes and onions, followed by a tinned peach trifle and coffee made in a percolator. I was staggered at how Wendy, who had never cooked in her life, managed it all. I could not believe that Hamisi had made any major contribution, as I knew from bitter experience that his *specialité de la maison* was fried antelope brains. Conversation was punctuated by the quarterly Westminster chimes emanating from Noel's clock. (Fifty years later the clock still chimes in our kitchen.) After dinner we played auction bridge.

One Sunday, when the Robinsons from Kondoa were staying with Noel, we all set off in three cars for a picnic. Having visited George Rex's grave at Galappo, we lunched on the western slopes of Ufiome Mountain where we overlooked Babati Lake with the Great Rift Wall in the distance. Then we had tea on top of the escarpment, 6,000 feet above sea level, looking towards Kikori and out to the Masai steppe and seeing all the country that I had tramped over. We sat on rocks amidst ferns and the lovely feathery leafed *Brachystegia* trees. It was the cold season and very pleasant.

The mail was extremely irregular, mainly due to faults at our end. My record was a letter from England which took 10 weeks to reach me. It had "Postage Due" on it for some small amount, but the postal clerk at Kondoa had no idea of what it all meant and so locked up all such letters in the safe. It was not until he had been sacked for inefficiency that his successor found the hoard and distributed the letters.

Wendy had so far seen no game, and was keen on going shooting with me on my usual Saturday morning meat-hunt. When I told Kompania that she would be coming with us he said lugubriously, "women always bring trouble."

We went to an open *mbuga* where there were usually giraffe and antelope, but nothing worse. For hours we waded through exceptionally high grass, the result of the previous season's abnormally heavy rainfall – 55 inches. To Wendy's delight we saw four giraffe. Shortly afterwards, Saidi, my local guide almost trod on a huge puff adder – so she saw that. In addition she saw sated vultures lingering round an old lion kill.

On our way home we walked along the edge of an extensive thicket where the grass was somewhat shorter and the going easier. I was rather worried to see rhino dung, but it was several days old; however, I took the precaution of changing over to hard-nosed bullets. Suddenly we saw fresh spoor and then heard the three sharp snorts of an alerted rhino; he was very close. I swung round and faced the thicket, awaiting the expected charge. Fortunately, the gentle wind must have veered slightly preventing the rhino from getting a confirmatory whiff of our scent, but he was still there as there was no crashing of branches. I seized Wendy by the hand and drew her away from the line of the wind, as quietly as possible, but then I had only one free hand and she was between me and the rhino – had it charged. (Wendy had no idea of what was going on, or what she was supposed to do under such circumstances.) Luckily we got away without trouble. Kompania was right – I never took Wendy hunting again.

During August the station was very empty with only ourselves and three bachelors – Potts, the Senior Entomologist, Scott and Moggridge. Mog, as he was always known was great fun: his favourite expression was 'I haven't laughed so much since father died' – incidentally his father, an admiral, was still very much alive. Mog and I once had a most ridiculous evening. Having found that tsetse did not normally attack man after dark, I wondered whether the presence of an ox would alter the picture. (I had managed to get several beasts with great difficulty.) Mog had volunteered to accompany me as my armed escort. We set off at dusk and had an hilarious time trying to get the ox to behave. He settled down, and we managed to catch several flies off him before a lion started grunting in our vicinity: the ox went berserk. The shouts, yells and curses, in two languages, must have terrified the lion, whereas we forgot its existence in our struggle not to lose our precious ox. Finally, we got home to drink some much needed whisky. The experiment was not repeated.

Car repairs were a great problem as there were no garages: I do not think there was one even in Dodoma, 150 miles away. A short-circuit developed in "Fat Fatima's" lighting system. Tschope was our only hope. On his next visit, I got him to look at her. He was an excellent mechanic, but no electrician: he blew all four lamps. We had no lights for six weeks, but then most fortunately Mog had a friend to stay, who kindly traced the fault.

Petrol had to be brought from Dodoma to Kondoa by lorry. It was sold by the case – a wooden box containing two, four gallon tin cans. Life in bush was dependant on both empty cases and cans.

The boxes were 21 inches long, 10 inches wide and 14½ inches deep. Two boxes on end with another laid horizontally across them made an excellent dressing table; the mouths faced outwards: the floor of the horizontal box held hair-brushes, razor etc., and the floors of the vertical boxes held shoes. Boxes stacked horizontally could be used for books or for office files.

The boxes could be modified by affixing hinges to the original lids and hasps to take padlocks; the contents were indicated by a stencilled code e.g. COOK – his utensils, DRINK – filter etc., MILK – tins of, CHOP – tinned food, flour etc. Such boxes were kept ready for porter trek, and labelled so that on reaching the destination one could immediately get the desired item – usually DRINK. (I still have the MILK and CHOP boxes in which I now store potatoes, and a photograph of Burtt shaving, nude as to the waist and sitting on a petrol box inscribed – SPHINX. HIGHLY INFLAMMABLE).

The petrol tins were indispensable for bringing water to the camp and for heating a tin for the bath. The cook used two lidless cans, one on each side of the fire, as ovens, they were linked by a piece of expanded metal on top, on which he placed his saucepans: the garden boy used them for watering the flowers and vegetables and I used them for breeding maggots for my parasite work, which will be referred to later.

The above is my *in memoriam* to the now defunct petrol case and its tins, which have been replaced by filling stations and 200 litre drums.

110

On my return from leave I gave Musa and Hamisi my old gramophone; they were delighted and played Swahili dance records continuously which kept everyone cheerful. We also took on Arser, Musa's brother, as a second servant. I became suspicious that he was stealing my cigarettes, and so always hid them before going to bed. One night I forgot to do so, and next morning I could not find them anywhere. Finally I asked Arser if he had seen them. "Oh! Yes *bwana*. I realised that you had forgotten to hide them, so I hid them for you. Here they are."

The mud walls of Rhino Hall had become infested with white-ants (termites). They erupted in a corner of our dressing room, and every morning Arser would remove the mound of earth thrown up overnight. I tried everything I could think of from kerosene to boiling water, but the ants were quite unperturbed. (In those days there were very few efficient insecticides.) Wendy's pride and joy was her wardrobe trunk: it stood on end, open, with all her dresses hanging on coat hangers. For safety, we moved her wardrobe well away from the walls to a position where it got in everyone's way. One morning we found that overnight the white ants had built an earth tunnel across the floor and had started eating a skirt. We were both furious. I racked by brains and then remembered the fire extinguisher that had been sent out by the head office clerk in a fit of mental aberration; I, suffering from a similar fit, shovelled the earth away to expose the mouth of the burrow: I, who had never used a fire extinguisher and thought the spray came out of the nose, smote it on the ground and applied it to the burrow. Needless to say, the jet shot out from between my legs and drenched Hamisi, who was inquisitive and unbeknown to me had come to see the fun. That night I could not sleep for the termites' laughter.

The Robinsons came out from Kondoa to spend a weekend with us. All that Friday night and Saturday morning there was a great dance in Kikori village to celebrate the coming of age of the daughter of *Mwanangwa* Tanganeza – the village head. People had been invited from miles around.

We went down in the evening and on our arrival the headman stopped the drums. We were escorted into the enclosure and were then presented to the witch-doctor. He wore a khaki pith helmet and trousers: stork feathers were strapped all over him, and his stomach was padded out to a huge size with a bolster. (I thought that he might lay an egg at any moment.) We then had to parade round the throng of dancers and inspect them; all the dancers stood to salute and

111

bellowed out a chant. Having made a complete tour of inspection there was a blare from the bugle and I had to make a speech. The dance then continued until the arrival of the King.

There was a great flourish of instruments and then a small dignified figure wearing white ducks appeared. He wore a Sam Browne belt of blue cloth, a huge sword in scabbard and a crown cut out of cardboard from a shoe box. As the drums thumped and the bugle blared, he stood to attention answering the salute. Then with superb dignity, hand on sword pommel, he walked round the ranks – an exact parody of the Governor. It was our Musa!

The dance now started. There were two hide drums, a petrol tin and a bugle.

The crowds moved round and round, a slow rhythmic march in which arms, hands, head, feet – all kept the most perfect time. They kept time with every inch of their bodies. The men were all in one column and the women in another. The tiny children were among the best performers. Everyone wore a mask-like face.

Periodically an old man, who was almost naked, tore round the ring, then dipped an antelope tail into a gourd of honey and spattered the liquid on the feet of the dancers to purge them of all illness. The blood relations of the daughter who had come of age, smeared their faces with the sticky seeds of some plant.

At about 10.30 p.m. we signified our wish to depart, the dance stopped and the crowds were drawn apart to enable us to leave.

All night long the drums throbbed.

After breakfast we went down to see the Coming of Age Ceremony. The scene was gayer in daylight and not so weird. Periodically a body of male dancers detached themselves from the throng, and marched in single file round the village in search of enemies, and then returned to report "all clear".

For three years the headman's daughter had been kept hidden in a hut. At last her hour came, and the procession appeared.

First came an old man carrying the 'bride' on his shoulders with her legs around his neck: following her was her 'bridesmaid', similarly mounted. They were surrounded by a yelling crowd of relatives, one of whom held out a bowl for alms. The two girls were dressed in red and white calico and their faces were spotted with sticky seeds. It was the behaviour of the 'bride' which lent a bizarre effect to the procession. She made the flesh of her whole body shiver and wobble like a jelly: tremors seemed to sweep through her: she swayed from her hips as if about to swoon, and her eyes were tight shut, except for an occasional gleam from under the drooped lids.

Round and round she was carried, with people dropping alms into the bowl. Finally she was taken back to her house, the dancers danced, and we discreetly left.

Noel, Maria and a girl-friend were also out for the weekend. Thus there were three youthful women, and in the evening we went over to Potts' house and danced until midnight – the first, and I think the last, European dance to be held at Kikori.

An earth tennis court had recently been made, and I had my first game for three years; however, soon afterwards we heard that headquarters and five members of staff, including Noel, were to be transferred to Shinyanga – the centre for tsetse reclamation work – and some 300 miles from Kikori. Our social life was not to last much longer.

Two other bits of news were that due to the slump the length of a tour in East Africa was to be increased from 2 to 2½ years, and that it would be at least 2½ years before we were made pensionable.

Arser got into trouble by having an affair with a villager's wife in the husband's absence; he feared that this would mean two year's imprisonment. Fortunately, the injured husband was not vindictive as he made allowance for Arser's extreme youth. I told Arser that he had better get married quickly if he wanted to stay with me, so he started hunting desperately for the most beautiful girl available. A month later I gave Arser the day off to go and bargain for the chosen girl: the bride price was 200 shillings, but he was convincd that he would get her for sixty. He did.

A party of us spent an hilarious weekend at Lake Basotu. We drove our cars onto a promontory and formed a laager with them. We all spent the afternoon watching the hippo disporting themselves in the water quite close to the shore. It was the first time that Wendy and I had seen them in the wild.

In the evening we were sitting round the fire within the laager, having our first drinks, when we heard the hippo making a lot of noise. Noel and I decided to go down to the lake shore with a light to see what they were up to. They came out of the water like moths to a lamp, so we beat a hasty retreat and rejoined the party. The hippo came very close to our laager, so we doused our lamps and forgot about them. Suddenly one of us saw by the glow of the fire that we had been joined by dozens of field mice. Bill Potts, who was gnawing a sausage roll, suddenly said "Pouff": all the mice jumped. So we spent our time saying "Pouff", and watching the mice jump.

It was all too stupid. Wendy and I spent our first night sleeping in "Fat Fatima" and found her very comfortable.

Next morning, Bill and I were standing on a hill in a shimmering heat, and looking down into a valley. I said "look, there is a rhino," pointing to a large animal about a mile away. Bill, who had been there before, said that it was a wart-hog, so I bet him ten shillings that it wasn't. We walked towards it, and the nearer we got the smaller it became; it was a mirage and a wart-hog, so I had to pay up.

When we got home, we found lion spoor by our front door.

In mid-October the bush fires swept in from Masailand. They were on a colossal scale owing to the height of the grass after an abnormally heavy rainfall. All day the sky was pea-soup coloured and the crackling sounded like machine-gun fire. For 24 hours we never saw the sun.

All my staff, including the servants, had spent the morning burning firebreaks around our camp – back-burning into the wind, and when the break was wide enough, beating out the flames with branches lest our fire escape and destroy the village below us. We were dripping with sweat. (The hairs on my arms were singed, which Wendy said was a great improvement.)

It was as well that we had worked so hard, as eventually the fire swept up our valley at tremendous speed, fanned by a hurricane of wind. So strong was the gale that it produced horizontal flames some 30 feet in length; tall trees had their topmost leaves scorched and often a rotten bough caught fire. As the flames tore past one could hear a constant high-pitched squeaking in the grass from hundreds of mice and rats that were perishing – we had had a plague of them. All night long the sky was tinted. Behind us, distant fires on top of invisible hills appeared as red arcs set in the sky.

Next morning the bush was black and the air full of fine feathery ashes. It was a relief to have the bush swept of its matted, tick-infested grass and not to have to de-tick oneself on return from bush before they swelled up into obscene little bags of blood. (A tick's life cannot be a happy one – sitting on a blade of grass, hoping for a passer-by, and then after weeks of patient waiting – just getting burnt alive.)

Several days later I was stalking a buck from behind an acacia when Kompania yelled "look out bwana": the tree under which we were standing had started to fall. We ran like greyhounds and the top boughs hit the ground about two yards behind us. Examination revealed that the recent bush fire had set alight to the base of the trunk which had been smouldering for some days on the opposite side to that of our approach.

114

There was no wind, and hence no smoke. The fire had eaten right through, and hence the silent start to the tree's fall; it was eerie.

One day when I was going to do the old Kandaga fly-round I took Wendy with me. On that day of troubles, we first had to repair a large hole in a bridge which spanned a stream. On reaching Kandaga, I took Wendy to my now dilapidated Zebra Hall and left her there with Musa, whilst I carried out the fly-round.

I had reached a part of the round which traversed an extensive thicket when I heard a rustling in the dense bushes beside me; it sounded as if a huge python was slithering away. Being armed with only a fly-net, my gun bearer having fallen behind to relieve himself, I stood still as I did not want to annoy whatever creature was there. The rustling stopped and a magnificent lion bounded across a glade. I then found that it had been sleeping within two yards of where I had been standing when it woke up.

On our way back to Kikori we drove through torrential rain – the first of the season. On reaching home we found that the rain had flooded the verandah and that Arser had stacked the furniture in our tiny dining room. So we picnicked until bedtime.

I was putting out the kerosene lamps when I felt nips, like red-hot needles, on my legs. I looked down – the place was swarming with siafu ants. We tore into the dining room, slammed the door, and laid a trail of wood ash across the threshold. We finished that room only to find ants pouring through the bedroom window and the bathroom door. For an hour we spread wood ash, of which I kept a large supply in the house, along the weak points in our defences and killed all the ants we could find in the house. In desperation we stood the legs of the beds in tins of water, and retreated into our strongholds. By the next morning the siafu ants had disappeared, leaving us with a temporarily insect-free house.

Noel had started packing up the office in Kondoa for transference to Shinyanga when, in early December, he fell sick. There was no doctor on the station, but the Goan sub-assistant surgeon suspected paratyphoid. A blood sample was despatched by runner to Dodoma: the unfortunate man was killed and eaten by a lion en route: no trace of the blood sample was found amongst the scattered mail, so it was assumed that the lion had swallowed it. Noel was seriously ill for many weeks and a doctor came up from Dodoma to see him; the provisional diagnosis was correct.

Swynnerton had come back from leave soon after Noel fell sick and infuriated Maria by dropping in to see how he was getting on, and then discussing files with him for hours: Noel was running a

115

temperature of 102°F at the time. Maria's tribulations were relieved for a week when Swynnerton came to Kikori to see us. It was very good to see him again but we were exhausted by the end of his visit, with only ten days left before Christmas.

On Christmas Day we had a picnic lunch in a cave at Masange, some 25 miles south of Kikori; there were seven of us, with Wendy the only woman. Hanging from the roof of the cave were hundreds of stala*tites*, – how helpful the wit who devised the mnemonic – 'tights go down and mites go up'. Nearby a waterfall gurgled down the face of the escarpment, falling over natural steps in the rock up which we climbed to get a magnificent view of the Masai steppe.

Later we drove on to Itundwe where we were spending the night in St. Clair Thompson's camp; he was a botanist and was growing plants inimical to tsetse fly. We were looking forward to a very cheeful party when Swynnerton arrived quite unexpectedly. He told us that he had motored up the 125 miles from Dodoma to wish us all a Happy Christmas. In reality he had come to talk tsetse flies which he did continuously until 3.00 a.m., by which time Wendy had been sound asleep in her deckchair for many hours.

We went to bed intending to sleep until 9.00 a.m. but at 5.30 a.m. Swynnerton called to his servant to bring his early morning tea and then started hammering on his typewriter. We were so exasperated that we told Thompson, our host, that we were setting off for home. He said, "There is only one day in the year when I don't want to talk 'tsetse fly' and that is Christmas Day, and of course Swyn. has to turn up and wreck everything." (The night before Swynnerton had asked me if I would drive him next day to Kikori and show him my parasite breeding work and where I intended to release them. I told him that I would not, that it would be Boxing Day, that I was always out in bush away from Wendy and that I was going to spend the holiday with her. He looked pained and said no more, but I felt a cad.)

There was a curious love/exasperation relationship between Swynnerton and his staff. We all liked and admired him – but he nearly drove us mad.

We got back to Kikori in good time, as we had the station coming to dinner – Potts and Moggridge! Wendy had gone to tremendous trouble decorating the house with artificial holly and mistletoe and writing out a menu with a frieze of rhino on the top. Our two guests tried to be cheerful but had eaten too much and slept too little to be more than drowsy, as were we ourselves. The evening was a flop.

116

Not having been to Kondoa for 10 weeks, one Friday we decided to visit Maria and Noel before they moved to Shinyanga, and to drive the long way round via Babati which was on the so-called Cape to Cairo road. On reaching this road I stopped to look at the car springs, which were making a horrible noise. At that moment Karl Neurk, a very nice Czechoslovakian settler whom I barely knew, pulled up and asked if he could help us. He diagnosed the trouble and insisted that we should come along to his farm where he had a workshop and an African mechanic. For the next two hours we were most hospitably entertained by our host and his friends – an Austrian Baron and a German.

The latest bit of gossip was that a German film company, operating from Babati, had lost an aeroplane. The pilot, flying only a few feet above ground level, was pursuing a rhino whilst the cameraman took pictures. Suddenly the aeroplane wing hit a white-ant heap, the plane stopped, the rhino turned and charged. The occupants got out just in time and watched the rhino stamping the remains of the machine into the ground. We saw the few salvaged parts outside Neurk's house.

We set off for Kondoa at 2 p.m. Just before ascending the escarpment by a long, steep, twisty road it started to rain. The road was like glass, with water pouring down it: on the near side was a cliff, on the off side a precipice. The car skidded from side to side: she climbed like a crab, but just made the top. A few miles later one of the chains broke. Finally we reached the river which bisects Kondoa; I asked if it was safe to cross and was assured that it was. It was just dark, as we slid down the bank into the river: on reaching the middle the engine stopped and the car sank down into a quicksand. We were bogged with the water rippling by. Every moment's delay was dangerous as after heavy rain the river came down in spate; only the year before Tschope's lorry had been swept away by a wall of water. I carried Wendy to the far bank and sent Musa to Grant, the District Officer, for help. For two and a half hours we struggled by torch light gaining inch by inch, until hard sand was reached and with engine roaring "Fat Fatima" ploughed her way up the far bank.

It had been the sort of day when things went wrong, but typical of the way we all helped each other in adversity and of how unexpected visitors knew that they would get a warm welcome, as we did at 8.30 p.m. from Noel and Maria who had no idea that we were coming in to spend the weekend. With communications so bad one expected the unexpected.

117

Noel was very cheerful, but as thin as a skeleton after his paratyphoid. Phillips had got back from leave the day before, but had left his wife and children behind in South Africa, so Wendy would continue to be the only woman in Kikori.

Life at Kikori was very dull for a solitary woman, more especially now that Noel's excellent library had gone with him to Shinyanga. On first hearing news of his transfer we all joined The Times Book Club and each got one book monthly, but as the number of staff fell from 7 to 2, so did the number of books.

Wendy took to dress making in a big way. She bought Japanese silks from the Indian shop in Kondoa and made herself evening dresses at about 15/- each. The women of Kondoa must have thought her terribly extravagant with her latest Paris creations. They fitted perfectly. In the absence of a long mirror, I suspected that she borrowed my shaving mirror.

On Sundays I used to cut her hair. Once I cut her ear, not with the blades but from behind the pivot, so intent was I on watching the points. I told her that she should not have such wee ears all hidden by fur. Wendy used to cut my hair: the use of thinning out scissors overcame the ridge and furrow effect produced by Musa in my bachelor days.

It was the short dry season – a month's gap between the early and heavy rains. The village mangoes and custard apples were ripe.

Babati had recently been made a subdivision of Mbulu District with its own A.D.O. – Graham de Courcy-Ireland by name. He had asked us over to lunch one Sunday and we spent a delightful day with him. He had built his *boma* of mud and wattle houses on the western slopes of Ufiome Mountain, which towered up behind him. In front and far below lay Babati Lake and beyond that the Rift Wall, with Hanang Mountain some 30 miles away. A streamlet trickled past his house, so he had the noise of running water. The site he had chosen was wonderfully cool.

In March 1931 we got two air mail letters, the first we had ever received: the first, from Egypt, took 17 days and the second, from England, 20 days, but we hoped that they would be quicker in the dry season.

In the same month we both got influenza, a rare infection in the days before air travel. In 1919 however, there had been a terrible epidemic in East Africa, which killed thousands of natives. The villagers, realising that the disease was highly infectious, put the dying outside their huts where the hyaenas found them, took them

118

away and ate them. The hyaenas, finding that such people put up no struggle, lost their fear of man, and started attacking women and children in daylight when they left the village to get water or firewood. I believe it was quite a problem eliminating the man-eating hyaenas. Even in my time if one was engaging a boy and asked him his age he might reply "I was born two years after the Green Vomit."

Victor Findlay returned from leave, having spent a few days with my brother in Egypt, and brought us presents from him. My Turkish cigarettes were smokeable, but Wendy's Turkish Delight was uneatable, having been packed near camphor – so she smoked my cigarettes.

An epidemic of fowl typhoid swept through the area and killed 37 of my 40 birds: eggs and chickens were locally unobtainable. Game was scarce and hard to find in the long grass. Even the disgusting mud fish were uncatchable: there was too much water in the stream. We lived on anything that could be made of flour. Then I managed to buy a flock of sheep – 4 to be exact – from the top of the escarpment which was tsetse-free; we would kill one, then bolt it before it went bad, and then kill and eat the others before they became emaciated and died of *nagana*. Indigestion was inevitable.

One day, in the heavy rains, we had to go into Kondoa. The road was ghastly, but "Fat Fatima", being a Ford, did the 50 miles in 6 hours. In many places the eight foot high grass had been beaten down by rain and lay across the track so that one could not see the surface or the wart-hog holes; one steered by feel. In some places the grass heads from each side of the track interlocked and one had to plough through them; in others, there were stretches of mud which had to be charged with the engine roaring, the car skidding and swaying from side to side, and bucking like a horse. On the hills, storm-water had eroded deep gullies down the centre of the track; one had to drive with the wheels on each side and pray that the trench would not become too wide.

We returned two days later, after a further three inches of rain had fallen. When 12 miles from Kikori, I saw a pond across the road – some 60 yards wide, with a 'river' swirling down the centre. (It had not been there two days before.) Kompania went ahead to measure the depth; it was knee-deep in the centre, so we might just manage. As it was pouring with rain and the 'river' was bound to get deeper, I charged it. We roared through until we reached the middle when the engine died: the fan had thrown water onto the plugs. Water was also

119

flowing through the car bonnet. Leaving Wendy in splendid isolation, I tied a rope to the car, and waded across. By the time I reached the other side the crowd of onlookers had disappeared; however, I had some old friends in a nearby shamba, and the owner, his wives and his children came and helped to pull us out. Despite her wetting, the car started at once and we got safely home.

Under such conditions, driving is an acquired art: one keeps one's eyes several yards ahead of the car and travels at about 5 m.p.h. constantly moving the steering wheel to cut each hump or trough diagonally, and so reduce the bump. On first getting home on leave it was difficult to remember to raise your eye-level to a more reasonable distance ahead. I drove in Africa, Wendy drove at home.

On such dreadful roads one was constantly breaking springs. The Ford had transverse springs, which were very difficult to stretch from one shackle-pin to the other. When fitting a rear spring, having got one end in, one had to put a car jack on the differential of the rear axle and stretch the spring to get the other end in. Something always slipped and left me with bloody knuckles – but I never was mechanically minded. One always carried a block of wood to wire onto the front axle, when a front spring broke; this enabled one to limp home. One also carried a tin of Colman's mustard to pour into the radiator if it leaked, or in tsetse-free country one could rely on donkey or horse manure to serve the same purpose. But in those days cars were relatively simple, and even I could clean a Zenith carburettor; further, the car instruction books were written for dumb clots like myself – not that they mentioned mustard and donkey dung.

I was very proud of my garden. From the village one saw a little white house on a hill nestling in a blaze of colour which merged into the bush. It merged – because when I had seeds I would get a bit more bush dug up and chuck seeds in, next morning I might see antelope or hyaena spoor on the freshly dug soil. Musa, Arser and Hamisi hated lending me a hand in the garden, but I always told them that they were as good as working for themselves, as when they died I would have flowers to put on their graves (hoots of mirth). They never could understand why Europeans waste time growing inedible flowers, nor could I really explain.

I had just taken Wendy into Kondoa to stay with the Robinsons, who were kindly putting her up whilst I camped out in bush, when she developed suspected amoebic dysentry. As the Goan sub-assistant surgeon could not reach a conclusion, I took a specimen to

120

the doctor in Dodoma who confirmed my fears and gave me detailed instructions for her treatment. Fortunately the rains were almost over, so the 200 miles return consultation presented no difficulty. It was a mild attack, she soon recovered and never had the disease again.

News came through that both Phillips* and Scott would soon be returning to South Africa for good, and that we were to live in Phillips' relatively palatial house. I owed him a lot for his ecological teaching, which was to prove invaluable to me in my future research. They sailed at the end of June, 1931.

*He had accepted a chair at Witwatersrand University.

CHAPTER 10

A Change of House

Although only two miles to the south our new house was in a very different setting. It was sited within *miombo* woodland on a gently sloping hill rising to the top of the escarpment. It was in bush and not overlooking the village, which from the health point of view was an improvement, but we were to miss our old view. In the dry season the woodland was frequented by lion, so much so that the occupants of the prefabricated houses gave up sleeping in their open verandahs. (These houses were dismantled, transported by our huge Guy lorry, and reassembled at Shinyanga.) Wendy and I would sit on the verandah steps in the evenings and sip our whiskies whilst listening to the lions grunting, and sometimes giving tongue, as they walked along the motorable track to drink at our stream, some 200 yards below us. They were very gentle lions and gave no trouble.

The Guy lorry referred to above, was ordered by Swynnerton whilst on leave; its abilities had been demonstrated to him in a hazel copse. It had two mechanically operated saws in front which cut down the saplings, and a winch behind, which pulled the larger trees out by the roots. Swynnerton thought that the Guy would be ideal for clearing bush in tsetse eradication schemes. On its arrival the Public Works Department complained that it was too heavy for their bridges, but somehow Swynnerton overcame that problem. It was the dry season when the lorry reached Kikori, and Swynnerton was keen to demonstrate its abilities. It was not too good when sawing through our fire resistant thickets, but when it came to winching a tree out of the rock-hard soil its performance was pathetic: the lorry strained and heaved and then started to be drawn backwards by the immovable tree. An axeman then cut the tree a quarter way through – no good, and then halfway through, and only then did the Guy's pull break the tree. At last the Guy lorry had a job which it could do – transporting houses.

I extended the clearing round the Phillips house to let in more breeze and give us a better view. The ground was rocky and the soil very shallow which complicated gardening.

We made a number of alterations to the house. Our predecessor disliked being disturbed, and had his kitchen some 60 yards from the house and the pit latrine nearly a hundred. We, liking the chatter of the boys and disliking even wetter soup and long journeys in the heavy rains, changed all this. We built a mud kitchen near the house and cut off a section of the bathroom to make an indoor lavatory with a trapdoor to the outside for removal of the bucket. (In a similar setting in Kondoa and whilst staying with the Robinsons, when Wendy was ready to leave the lavatory she saw a snake curled round Robin's dressing gown which was hanging on the door; in response to her yells we removed the bucket and Wendy made an undignified exit into the garden. She was prone to such accidents: two years later in Nigeria, Wendy, the only lady at a dinner party given prior to a club dance, gracefully retired at the end of the meal: soon there was a banging on the door: she could not undo the lock: her host and I engineered her escape, I holding up her long silk dress. She rejoined the party, all young army officers, and all pretended not to have noticed her absence.)

Determined to combat fly-carried dysentery and mosquito-borne malaria, we netted the whole house and had the food kept in a pantry with a hatch into the dining room. It was a two-bedroomed house which was to prove most valuable.

We had the house whitewashed and the doors and windows painted black. At the weekend I cut out a rhino stencil and made a procession of rhinos going round the verandah walls and ending at a waterhole. The heads were polished and hung and Wendy's curtains put up. We moved into Rhino Hall III a fortnight after Phillips had left. The reason for the rush was that visitors were expected.

Hornby, the Veterinary Pathologist from Mpwapwa, arrived with his family. He wanted to see the research work and in particular some cattle which had been injected with tartar emetic, in an effort to protect them from *nagana* in an area where tsetse flies swarmed.

Shortly afterwards, I got a message that Swynnerton was flying from Shinyanga to the Babati landing strip where I was to meet him next day. We got there at 8.30 a.m. and waited until 6.00 p.m. but no plane arrived so we returned home. Next day I had just got back from the office when Swynnerton arrived by road; after 10 minutes for lunch, I drove him to Babati where he wanted to see the Governor, Sir Stewart Symes. We got there at 2.00 p.m. and waited by the roadside, with the Babati officials. Incidentally, Babati was in Mbulu District, not Kondoa. The road was lined by the Wa-ufiome tribesmen, armed with their spears and shields and naked except for a skin. During the long wait they kept up a song, stamping on the ground with their feet, and beating their shields with their spears. H.E. arrived at 5.30 p.m: we left Swynnerton and bolted, as Wendy and I were very untidy, having had no prior warning. Swynnerton returned and announced that H.E. was visiting Kikori next morning.

We duly met the Governor in bush, some miles from Kikori, where I showed him a tsetse rock-breeding site, where the fly gives birth to a larva, and then on to my laboratory. He was particularly interested in the parasite breeding work. Sir Stewart was a most charming man, every inch a soldier, lean, brown and athletic. We then went to our house where Wendy had chocolate cake and sandwiches awaiting him: they had a long talk: he then went on to see Pott's work, and finally left at 2.00 p.m. (He had refused lunch which Wendy had prepared; he disliked inflicting his large retinue on people, so they always carried their own food.)

Our next visitors were the Middletons, who spent three nights with us. He was the doctor, recently posted to Kondoa; we were more than delighted, as we thought that there was a baby on the way.

One day, Graham de Courcy-Ireland, who was a good friend of ours, came over from Babati, spent the day with us, and returned that evening. Next morning the natives asked him to come quickly as an elephant was devastating their crops. He had to refuse as Greening, his District Officer, was due to arrive from Mbulu on a visit of inspection. Soon Graham heard the elephant rampaging and the shouts of the natives. He felt that he had to go and help them. Apparently he fired four shots at the elephant's head. The infuriated beast charged down the hill and through the middle of the compound. Greening and his wife had just arrived and shot into the doorway of the house. As Graham followed in hot pursuit, Greening yelled to him not to use the head – but the shoulder shot. Greening, being unarmed, did not follow him.

Gordon-Russel, a friend of ours, was in his tent by the roadside. He heard a few shots and paid no attention. He then heard an elephant scream, and knew that it must have killed someone. He dashed out and spent a long time searching for the body in terribly thick jungle, with the maddened elephant charging here, there and

124

everywhere. When he found the body it was obvious that poor Graham had been killed at once. Apparently he had walked straight into the wounded beast, which had caught him with its trunk, smashed him onto the ground and transfixed him with its tusks, and then screamed and trumpeted.

The shock was so great that Gordon-Russell could not shoot as he was trembling too much. By this time the news had spread and some settlers were standing on the road; after they had fired many shots, a young *Afrikaner* climbed a tree, got a good view of the elephant and killed it.

When the news trickled through to us we just could not believe it: he had been so cheerful so short a time before: he was only about 23 years old, was such a charming fellow and due for leave in only a month's time.

It was to be 40 years before I again talked to someone with that uncommon surname. My nephew 'phoned me:

'Uncle Tam. I am engaged to be married.'

'Hearty congratulations. What is her name?'

'Rosamond de Courcy-Ireland.'

'Had she any relation who was killed by an elephant in Tanganyka?'

'I don't know, but she is sitting by me and I will hand over to her.'

After a few appropriate sentences, I repeated my question. She replied, 'Yes! he was my uncle.'

Potts accompanied us on a memorable Bank Holiday weekend, on a second more leisurely visit to Lake Basotu. On the Saturday we first motored westwards from Babati to the Rift Wall. On the way we met Wa-ufiome women wearing their ruffs of concentric circles of highly polished brass wire around their necks, and later the Wambulu women with their soft leather shawls beautifully decorated with beads, shells and coins; in some cases, the shawls contracted at the back of the waist and then widened out near ground level, looking like the tails of birds.

We then drove south-westwards along the foot of the Rift Wall and at one point scrambled up a steep path and looked along the Rift Valley to see the great salt lakes and solitary Hanang Mountain soaring up out of the plain to achieve a height of some 11,000 feet. We drove round the eastern and southern faces of the mountain and continued westwards through the Mangati plains where game abounded. Within some 25 yards of the track we saw zebra, giraffe, wildebeest, impala, Thomson's and Grant's gazelles, ostrich, secretary birds – which eat snakes, greater and lesser bustards and flocks of guinea fowl. It was like being in a zoo without bars. Occasionally we would see a Mangati herdsman grazing his cattle amongst the game.

We had driven nearly 100 miles by the time we reached Lake Basotu. We drove out onto a narrow promontory surrounded by water on three sides. As the car stopped we heard the snorting of hippo. There, in front of us, were hippo floating on the surface of the water, diving without a sound, and reappearing a minute later with a snort. There were six grown-ups and two baby hippo. These monstrous animals were delightful to watch: their ears were constantly flicking, and their eyes protruded on small, thick lumps of pink flesh. Every now and then one would pop out of the water showing his back, his open pink mouth with huge teeth, and his pink belly; then with a splash he would flop back.

Camp was ready by dusk and we sat and watched the pelicans ponderously flapping back to the shore. Flocks of parrots flew home screeching over our heads, guinea fowl cackled in the bushes, and Egyptian geese 'honk-honked' as they flashed by. The fish eagles gave up work and swooped down to their special roosting trees. The storks got tired of standing on one leg, and the Bateleur eagle of uttering its deep bark. The wind dropped, there was a hush only broken by the endless rustling of legions of mice that stalked us from all sides – scurry two feet forward, a movement, scurry one foot back.

As darkness fell so our hurricane lamps showed up more brightly and in proportion the hippo family objected. The mouse symphony was shattered by the blasts, snorts, squeals and grunts of the outraged and curious hippopotamuses. Pandemonium broke out in the water just below us. Shut your eyes and picture Hell's pig farm at night. The noise got worse and finally we heard the hippo splashing in the shallows beneath us, voicing their indignation and resentment. Admittedly hippo are charmingly grotesque at 50 yards range in daylight, but at 20 yards on a pitch black night they entirely destroy one's appreciation of John Haig. Being strongly attracted to light, we feared that they might loom up out of the blackness and stampede through the drinks. Hence we decided to be discreet and so moved ourselves, our lights and our glasses to behind the car. Our discretion was appreciated: the hippo stopped being indignant, and their snorts became playful.

Wendy and I slept in the car. Our sleep was fitful. When the hippo were quiet, the mice were twittering. Bill Potts slept in his camp bed and never heard a thing.

126

The dawn was lovely. The deep, vivid yellow-green of the fever trees* became flushed with copper. A flimsy mist drifted over the lake. Thousands of birds woke up and whistled, hooted, screeched and cackled – according to their parentage. Black snake-necked cormorants stood with out-stretched wings on islands of weed, aping the German eagle. Divers, grebes and coots bobbed about like canoes in the wake of stately barges – the pelicans.

So another day and night passed in paradise – and Monday found us home, unwashed, grimed, tired and thankful to have the prospect of a night unaccompanied by porcine grunts.

In August 1931 life at Kikori became more and more depressing. Detailed plans had been made for my brother, stationed in Egypt, to stay with us and then accompany us on my local leave which was to be spent with cousins in Kenya; he failed to materialise owing to a muddle between a shipping agent and the captain of a cargo boat. Potts and Moggridge were to leave Kikori for good on transfer to Shinyanga; only Victor Findlay would be left. The last straw was that owing to retrenchment due to the slump, Kondoa was also to be depleted; it would no longer have a doctor and the only Europeans left would be the D.O. and the A.D.O. The loss of a doctor was particularly worrying as the baby was expected in April, the height of the rains, when the 150 miles of road to Dodoma could become virtually impassable. We felt like the ship's cats when they see the last life-boat leave the sinking wreck.

One night the lions were very noisy as they tried to stampede our experimental cattle in the *kraal* below the house. We were used to our lions roaring but not the lions *and* bellowing oxen; the combination gave us a disturbed night. The only other event at this time was the shooting of a wart-hog behind the house, between the porridge and bacon-and-eggs course; the pork was tough – but a change: we had no fridge, so it had to be eaten quickly.

On return from leave, a new official was posted to Kondoa. He was the biggest snob I have ever met, and boasted that he never entertained anyone who had not been to either the University of Oxford or Cambridge. In consequence we had never been asked to his house. Then his wife came out – a most charming woman; she and Wendy liked each other. We were asked to dine with them. He began talking loftily about the partridge and pheasant shooting he

*Acacia xanthophloea

127

had had on leave, and how he was always asked to shoot by the Cooks, friends of his who live in a beautiful place called Windhurst. (I pricked up my ears and listened intently.) He went on to say how Robbie Cook had refused his advice to place a gun at a certain spot, and how the pheasants had streamed over it, and finished by saying – 'I suppose you have never done any shooting.' Having let him patronise me for ten minutes I replied 'I am delighted that my cousin was able to give you such good shooting.' He never did again.

CHAPTER 11

A Visit to Kenya

I was due for 3 weeks local leave and had arranged for us to visit my first cousin Evelyn, and her husband Tim Brodhurst-Hill, in Kenya. After the war he had been given land under the Soldier Settlement Scheme and was growing coffee at KipKarren, near Turbo, and some 30 miles west of Eldoret.

Unfortunately, my letters about the Kenya visit have gone astray so I must rely on memory and a photograph album. We set off on September 10, 1931, on the Cape to Cairo road for Arusha. (I first used this road in early 1928 when I and my porters walked the last 30 miles into Kondoa. We had not seen a car all day, when suddenly one came bumping along, stopped, and an American said 'Say! Is this right for Cape Town?' I hadn't a clue – the thought that such an awful road could go such an awful distance was inconceivable. He put his foot on the gas and buzzed off.)

We spent the first night in Arusha, at Paynter's Hotel. On a wall hung a mudguard, punctured by the teeth of a lion in its death snap; it had been shot as it charged the car.

Next morning we set off by the Great North Road for Kenya, skirting Mt. Meru. For miles we crawled through deep volcanic ash leaving an impenetrable fog swirling behind us; the track was wide, as each driver thought he could find an easier route. On gentle dips the trouble was caused by invisible gullies, several feet wide and a foot deep, filled with soft dust. Several times the back axle stuck on a ridge between gullies, when we had to put down short planks on which to back out. Fortunately the Ford was a very powerful car with a high clearance. (In those days there was no English car so well suited to such bad roads.) We spent the night at the foot of Longido Mountain, where I saw a gerenuk standing on its hind legs reaching for branches.

Next day we entered Kenya at Namanga where an Indian ghee

(clarified butter) factory had recently been wrecked by elephants; as this was Masai country, there was much milk for sale. We ran into trouble in the lonely Athi plains where there had been heavy rain; frequently we had to dig ourselves out of the clinging black cotton soil. It was dark by the time we reached the Ngong hills where we got lost. No signpost ever mentioned Nairobi; it was too well known. Eventually we reached the New Stanley Hotel; as we drove up the car's steering died and we slithered to an ungainly stop.

After several days we set off for Bahati, north of Nakuru, where we were going to spend a couple of nights with Michael Blundell (later Sir Michael). We had been at Wellington College together, and by chance when he first went out to Kenya he had stayed with my cousins to study coffee growing. On the drive up-country, we deviated to look at the pink carpets of flamingoes on Lakes Elementeita and Nakuru; on the latter lake Huxley estimated that these birds could be numbered by the hundred thousand. On leaving Nakuru for Bahati, after climbing for some miles, we reached the lip of the Menengai crater, a perfect circle, some 7 miles in diameter, with a cone rising up from the centre. The floor was hundreds of feet below us, and we looked down on the backs of eagles hovering over the cedar* forest and slag heaps. It was a fantastic sight.

Michael Blundell was very kind to us and we enjoyed our brief visit. I was much impressed by his ability to speak two native languages fluently, which enabled him to get labour from the reserves when others failed. The next morning he was up at 5 a.m. cementing a tank for coffee beans. His house must have been at an altitude comparable to that of the crater top, as I have an early morning photograph, taken from his garden, with mist or steam billowing out of the crater.

Some miles after leaving Blundell's farm we ran into a cloud of locusts whirring around us, hitting the windscreen and entering by the glassless windows; one settled on my bare knee and revealed a remarkable grip as I plucked it off. Soon after driving out of the swarm, the engine stopped: the plugs were clean: the petrol was entering and leaving the carburettor and its filters were clean. I was scratching my head in perplexity when I noticed two bits of 'grass' protruding from the carburettor's air intake; investigation revealed the hind legs of a large locust, after whose careful removal the engine started up and all was well.

We started to climb; on reaching the top of each hill "Fat Fatima"

*So called in East Africa; it refers to *Podocarpus gracilior* – an indigenous timber tree.

130

boiled and had a rest until it was safe to undo the radiator cap and give her a long drink: she would then sail down to the bottom where there was always a stream from which to refill her water cans. Eventually we reached the highest point on the Cape to Cairo road, at 9,000 feet, and crossed the Equator in a bamboo forest – a setting which bore no resemblances to "Crossing the Line" in the ship's swimming pool when Father Neptune ducked the newcomers.

The road dropped down to Timboroa, a mere 7,000 feet, where there was a cedar forest, and thence to Burnt Forest – grassland out of which protruded the charred and blackened trunks of long-dead trees. Finally we drove through the small town of Eldoret and across the Uasin Gishu Plateau to the Kipkarren farmsteads. We had driven 1,000 miles since leaving Kikori.

It was in 1920, only eleven years before our visit, that my cousins went out to settle in Kenya. The railway only reached Londiani so they had to complete the last 63 miles to Eldoret by post-carts, each drawn by eight trotting oxen. But those and subsequent years have already been described by Evelyn in her two books – "The Youngest Lion" under the pseudonym of Eve Bache, and "So This is Kenya" under her own name, Brodhurst-Hill.

We found them living in a nice thatched farmhouse with an outbuilding for visitors, surrounded by a most attractive garden, and beyond, some acres of coffee.

Evelyn, who has been dead many years, was much older than myself and, although great fun, had a distinct sense of propriety. In the garden there were two, well separated little huts, heavily screened by shrubs, and each housing a pit latrine, one for Ladies and one for Gentlemen. One morning Wendy came back to our room with her eyes popping out of her head and said, "There is a large cobra by the lavatory seat." Feeling that under such circumstances a gentleman might enter the Ladies, I grabbed my shot gun and went to investigate. True enough – there was a large cobra, coiled, with head raised. I fired. There was a cloud of dust from the mud floor and walls. I could see nothing. I waited for the dust to settle. There was no snake! I entered very cautiously and studied the low thatched roof – no snake. I looked behind the seat – no snake. I looked into the pit. There was the cobra swimming around – a release of tension – I pulled the trigger – I was plastered. One of the onlookers who had gathered around ran to Evelyn and said, "A spitting cobra has spat in your 'brother's' face; he is blinded." In no time, Evelyn came running out with a basin of water and soon cleaned me up. "Tam,"

131

she said, "don't tell anyone about this, it would be terrible if our neighbours got to hear about it."

Evelyn, a parson's daughter, had a strong sense of duty towards her employees, and daily went round the staff quarters ministering to the sick wives and children. They loved her dearly, but I was glad that I was not one of her patients: her remedy for coughs was a spoonful of kerosene in milk, and her patients were convinced of its efficacy.

Soon after our arrival, I was asked to shoot a crop-raiding waterbuck. The climate was much colder than Kikori, and I went out inadequately dressed. A few days later, I developed a temperature and cough, and thinking it was malaria, took 30 grains of quinine daily and stayed in bed. After a week or more, I was feeling very ill, so Wendy drove me in to see Dr. Forbes in Eldoret. He diagnosed bronchial pneumonia, and hospitalised me. Next morning my temperature was normal; I had had the crisis at Kipkarren. I was kept in hospital for 12 days and Dr. Forbes insisted that I drank a jug of fresh orange juice daily – an improvement on kerosene and milk.

Meanwhile, Wendy had been staying with the Lesters, friends of Evelyn's, who farmed outside Eldoret; I joined them there to convalesce. They were the kindest of people. We had never met them before, yet they extended unlimited hospitality – so typical of East Africa. Bob Lester had the neatest of farms. He grew his own eucalyptus and wattle; it seemed so extraordinary to see *straight* poles, posts and stakes used for outbuildings and cattle kraals after the contorted bush timber I was used to.

We set off for home, and spent a night at the Avenue Hotel where the first lift in Nairobi had recently been installed. We took a different route back as we wanted to get a close view of Mt. Kilimanjaro. On the way there, we were going through thicket when the car stuck in a small sand drift. I had just reversed out of it when Wendy yelled, "Look." Some seven yards from her was a huge elephant with his ears raised. With horn blaring, I shot round the drift and, in the words of Tschope, "I put foot on zee gas and I go like zee bloody-hell."

On skirting the north face of Kilimanjaro we again went through miles of volcanic ash, and then climbed up the lower slopes through endless banana plantations – the staple food of the Wachagga tribe – and passed many mission stations with tinkling bells, to reach a little hotel at Marangu. (It was the only hotel I had ever been to where the proprietor advised me to take out the car battery, lest it be stolen overnight.)

132

As we sat on the verandah quenching our thirst and absorbing the beauty of the mountain's two peaks – snow capped Kibo and brown-topped Mawenzi – my mind wandered to that momentary vision I had had of Kibo at sunset, when camped 130 miles away on a volcanic cone below Ufiome Mountain. The vision had become a reality; how huge was reality.

CHAPTER 12

Back to Kikori:
Separation and Reunion

It was good to be home and to overhear Musa telling Hamisi of the great mountains he had seen, the strange tribes encountered and the story of *bwana* and the cobra – shrieks of mirth; such gossip went on for days.

During my absence, Heard had kindly looked after my colonies of tsetse parasites. They were kept in large glass-topped wooden boxes, with hinged lids and hasps, and stood on trestle benches; the boxes were housed in a shed with a thatched roof covered by a tarpaulin. Heard was an enthusiastic, amateur snake-handler. Shortly before our return, he noticed a snake's thin tail protruding from under the tarpaulin; he siezed it with his left hand and was slowly pulling it through his right, which was ready to grab it behind the head. However, it was not a little snake as he had envisaged, but a snake with an end but apparently no beginning. He then realised that it was the very poisonous mamba, but kept calm and let the coils encircle his legs. Finally, he grabbed it behind the head and yelled for help. Yussufu Cheke came to his assistance and Heard told him to fetch an empty parasite box. He did so, and bravely unwound the coils and pushed them into the box. Finally, Heard shoved the head in, slammed down the lid, and padlocked the hasps together. Very elated, he carried the box to his house and placed it in the doorway. Every time he passed, the snake struck at the glass. He then remembered the stories of how a mamba's mate often comes to look for its partner. He then decided to remove the snake's poison fangs and anaesthetized it with chloroform, but "most unfortunately he overdid it and the snake died" – which is the only part of the story which I disbelieve.

A few months later, Yussufu Cheke, who suffered from amoebic dysentery, became very ill and developed non-stop hiccups. I drove him into Kondoa, which took three hours; the Sub Assistant-

134

Surgeon diagnosed a liver abscess, and Yussufu died a few days later. He looked after the parasite breeding work, was most conscientious, and I was to miss him greatly.

I was breeding a tiny wasp-like insect which parasitises the pupae of both tsetse flies and the flesh-flies which lay their eggs on rotting meat; it was cheaper to rear the parasite on the latter. I had found that flesh-fly maggots were best reared on the heads of animals, shot for food, rather than on lumps of flesh which dried up. The heads were jammed into petrol tins, and suspended from the roof of a hut. One night we were awoken by the snarls, growls and screams of an enraged leopard who had got his head stuck in a tin. Off, up the hill he went, blinded and banging into trees. Next morning I followed the tracks and was relieved to find the tin – empty.

At dusk one evening, whilst strolling up to the house with Wendy and an empty shot gun, we saw a leopard standing on the path some 20 feet ahead of us. All three of us, equally startled, stood and stared – until I upped the gun and broke the spell.

The end of 1931 was a most depressing time. I had a large bill to pay for my hospitalisation in Kenya: owing to the slump a levy was imposed on our salaries, and allowances were stopped or reduced: there were staff cuts and Heard departed, leaving only ourselves and Victor in the Kikori area: then, there were rumours that our Department would be closed down in September 1932, unless the grant from the Colonial Development and Welfare Fund was renewed. We also had our own personal problems. There was no doctor for 150 miles, the baby was due towards the end of the heavy rains when the roads were almost impassable, and so we reluctantly decided that Wendy would have to go home for the event.

There was one bright spot, Cecil Stiebel who I had met on a train in Dar es Salaam, p. 97, replaced Robinson as A.D.O. Kondoa. We got to know him well – he was already a legendary character in the territory. His father, a South African and Provincial Commissioner, was under the impression that his son was studying medicine in London. He asked a friend who was going on leave to look the boy up. He did so at 11 a.m. one morning, but the landlady refused him admission as Mr. Stiebl was 'ill in bed.' Having explained the position, he virtually forced his way in to find Cecil on the bed, in full evening dress with opera hat on the floor, snoring his head off. It transpired tha Cecil had got bored with 'medicine' and was working for Mrs. Merrick, the night-club queen in the London of the nineteen-twenties. After this disclosure, his father got him into the

135

Colonial Service in Tanganyika Territory, where he could keep an eye on him.

Cecil was the collector of the most extraordinary items; his most macabre exhibit were 13 rupees, one on top of the other, fused down one side to form a solid block of metal; the owner had them tied up in a bit of cloth when he was struck by lightning. Cecil gave me some bronze coins which he had collected at Kisimani on Mafia Island; I had them identified – they were Persian coins – "Haroun al Kadia, circa 1450, of the WaaJemi tribe."

I was busy writing a lengthy scientific paper which was to earn me my D.Sc., but I could not write for long without getting severe headaches. In late January 1932 we went down to Dodoma whence the doctor despatched me to Dar es Salaam, some 270 miles, to get my eyes tested. (Friends kindly put Wendy up during my absence.) This visit to the coast turned out to be extremely fortunate. I spent two nights with Dr. and Mrs. Connell with whom I had travelled home on the boat. Bill Connell told me that it was absurd to contemplate sending Wendy home for the baby – what with seasickness and the separation, she would probably arrive a physical wreck: labour was usually easier in the tropics, possibly due to the taking of prophylactic quinine, and further in Dar hospital they had a specialist in Dr. Skelton M.D. (midwifery). What a relief it was to get expert advice: in Kondoa, at that time, there were only three men and not a single woman. My eye trouble had served us in good stead – I was suffering from tired eye muscles following on the pneumonia, rather than from defective vision.

We were most excited – for the first time the Royal Air Force was to 'show the flag' in Tanganyika Territory. They were to land at Kondoa on February 16th. Wendy cut my hair and in her excitement produced a moth-eaten effect, but daily she touched it up with Indian ink. I quote from my letter to my brother to recreate the atmosphere of those times.

'We went in to meet the R.A.F. on Tuesday. When still 15 miles from Kondoa we saw the planes and I said to Wendy 'Let's race them!" We had only done a quarter of a mile when they were almost out of sight. They then got lost, so we were first at the landing strip. We had just got out of the car when, with an awe inspiring roar, the four planes tore over the town, only 100 feet up, and flying wing to wing. My God! it was grand. The natives all bolted.'

'Well! we had a wonderful time with the lads and Wendy, being the only woman, was spoilt. The party consisted of Wing-

Commander Harris (later "Bomber Harris" and Marshall of the R.A.F.), Flight-Lieut. Atcherley of Schneider Trophy and King's Cup fame, Flying Officers Cooper and Jarman and six sergeants. They spent two days with us and life was one long rag. Kondoa, being such a tiny place, there was no etiquette.'

'On Wednesday afternoon the civilians were to be taken up for a flight, but not Wendy. We juniors had lunch with Cecil Stiebel, but lunch was late and we had to bolt a plate of curry and rush off to the air-strip. I then went for my flip – my first ever. I sat on a kerosene tin in the open back of the cockpit and was warned to take off my shirt lest the collar points flapped holes in my jowls; it certainly was breezy. I revelled in the first half hour, but the air was very bumpy and I was then sick over the side; the curry disappeared horizontally.'

'In one and a half hours, sometimes flying at 150 m.p.h., I saw much of the country over which I had sweated with blistered feet. Then Cooper, the navigator, got lost as there are no accurate maps of the country. If it hadn't been for poor old foot-slogging Tam, we would probably have landed in Zanzibar. In Tanganyika, aircraft tended to follow the roads, but that would have been too plebeian for the R.A.F.'

'Jarman, the pilot, did some stunts on my behalf; as he said later – "much better to be really sick, and not retch." This explained why, whenever I tried to take a photograph of the ground below me, I saw only sky in the camera's view-finder. At one point we swooped down from 9,000 feet onto the Dodoma road, did several miles with the wheels barely off the ground, and then zoomed up the Kirema Gorge with great cliffs on either side – up and over the top, scattering sheep, goats, and herdsmen – and then to soar above them. It was thrilling. We landed to find Wendy waiting for us.'

They *were* a nice crowd, those young pilots. I was much impressed by the way that they could never be persuaded to take more than one alcoholic drink, and mainly accepted soft drinks only. They had their mascot with them – a white bull-terrier. Before leaving, they gave us two superb photographs, one of the flight taken before their departure from Cairo with the pyramids far below them, and the other a group photograph of the 10 members of the flight with the bull-terrier at the feet of Wing-Commander Harris.

Our depression was considerably reduced by the R.A.F. visit. Soon afterwards the sun came out: we got news that the U.K. Government had renewed our Department's grant for the next 5 years, so that we were safe from retrenchment by the local authorities.

At the end of February the rains broke without warning. I rushed Wendy and Musa down the 150 miles to Dodoma; we had a rather alarming journey as the road surface was very skiddy. Next morning I saw them off on the train for Mpwapwa, where the Hornbys were very kindly putting her up for a week: she was then to go to other people in Dar es Salaam whom I barely knew. The baby was not due until early May, but I had got her out of the hinterland only just in time: we had 12½ inches of rain in March.

It was lonely back at Kikori, just trees and grass, grass and trees – one dreamt on, and the rest of the world seemed rather a fable and hardly likely to exist. At nights, Victor and I played poker-patience. We had a bad scare that sleeping sickness had broken out at Kikori, but it came to nothing. A native at Kandaga went mad: he decided to live in bush and shoot everyone he saw but sport was poor and he finally decided to shoot up his own villagers; they killed him and the owner of the spear got 6 months imprisonment, which seemed hard luck.

I got a letter from Wendy. She was very unhappy with the people she was staying with in Dar es Salaam; she sounded quite miserable. Who could I get to whisk her away? I knew so few people in Dar. Then I remembered a very tenuous connection. On my first voyage out I had been given a letter of introduction to Colonel and Mrs. Case – he was Commanding Officer of the King's African Rifles. The letter was from the Principal of a girls' school attended by my cousins and the Cases' daughter. The Cases were very senior people, but I wrote to Mrs. Case and explained the situation. On getting my letter she went round in her chauffeur-driven car and very tactfully removed Wendy and treated her like her own daughter: I had already arranged with Dr. and Mrs. Connell to have Wendy when the baby was near to arrival. How kind everyone was!

Fate was smiling on us. Wendy had heard of an English nurse who had left America to come and live with her brother who was prospecting in Tanganyika. On landing in Dar, she found that he had left for home on the previous boat having been deported as a Distressed British Subject. The girl, Katherine, was penniless, but adventurous, and was delighted to come up country with Wendy and help look after the baby.

On April 2nd I was at Dodoma and on my way to Shinyanga for a 12 day visit to our headquarters; I was delighted, as for 12 days I would be able to get in contact with Wendy, if need be.

I detrained at Shinyanga where a lorry was waiting to take me to our H.Q. at Old Shinyanga. We had gone some miles when we

138

reached a large stream that had come down in spate. Swynnerton and Noel were standing on the far bank. Swynnerton was very agitated as he was expecting an important visitor next day – Sir Sidney Armitage-Smith of the U.K. Treasury. A long, very narrow zinc bath was floated diagonally across on a rope. I was to be the guinea pig to see if it was stable – it wasn't. I got pulled half way across before it capsized and I had to swim for it. Swynnerton was very upset as he remembered that I had recently had pneumonia. They rushed me home to change and whilst doing so, got someone to solder the bung-holes of four empty petrol tins. We then returned to the stream, two tins were lashed to each side of the bath which was floated across; another guinea pig, with my suitcase, was successfully pulled back. I have in my album a photograph taken next morning of Sir Sidney doffing his sun helmet to Swynnerton whilst in mid-stream; the photo is irreverently labelled "Sir Sidney in bath."

Swynnerton, knowing that the visitor had come to see how he could cut our funds, had gone to great pains to make the visit a success. He had ascertained from down the line that Sir Sidney's favourite drink was gin and tonic and had managed to get the latter from Musoma. Before lunch Sir Sidney was appreciatively sipping his drink; he turned to his host and said, "You seem to do yourself very well Swynnerton." After that clanger, the Department's financial future depended on me. The night before, Sir Sidney had expressed a wish to shoot "a horn-ed beast" after lunch. Swynnerton had asked me to spend the morning looking for one, whilst he talked finance. Thanks to the local guides provided I found such a beast, and thanks to chance – it was still there after lunch and Sir Sidney did not miss.

During lunch two telegrams arrived. The first said, "Son born 8 a.m. both well" which I read out, the second from Musa said, "We have presented you with a fine male child" – which I kept to myself.

I was staying with Noel and Maria. After saying goodnight to the others we relaxed in their house. How lucky we had been! The baby was born a month earlier than expected; had Wendy gone home she would have been in a very late stage of pregnancy during the voyage.

After my Shinyanga visit, I investigated the islands in the south east part of Lake Victoria to see if conditions were suitable for the introduction of my parasite. I left Shinyanga by train at 4 a.m. for Mwanza in the Speke Gulf. At 7 p.m. I got aboard the Clement Hill. She was just like a small liner with spotlessly clean cabins, fans, hot and cold taps, smart English officers and Scots engineers. At dinner I sat next to the captain and ate English food. (It is extraordinary how sailors always turn their ships into a microcosm of their native land.) During the night we ran into a violent thunderstorm, it was very rough and I was nearly sick, since when I have classified the lake as an inland sea.

We reached Musoma at dawn, to find the entire European population of six people awaiting us; after the fourth gin and tonic we all had breakfast. One of the six inhabitants was Emson, the Veterinary Stock Inspector at Kondoa, who had been so kind to me in my first tour. I stayed with him and we had a great reunion.

I spent the next four days visiting various islands in canoes sewn up with string. Wherever the string goes through a hole in the wood, water enters as a small jet. Six paddlers paddle furiously chanting wild shanties, whilst the seventh crew-member bales madly – he is the most important man in the boat. When the island is three or four miles off-shore the crew know that speed is the essence of safety: it is a point of honour, *not* to fill up the holes. As these waters swarm with crocodiles, there is no point in being a good swimmer. Another hazard is presented by the waterspouts: one sees these gyrating columns of water, formed by a whirlwind between lake and cloud, buzzing about, but there are not many of them, they never can make up their minds as to where they are going, and I believe accidents are few.

The islands are clothed in wild sisal, each blade as sharp as a sword: in aloes, as prickly as hedgehogs: in milky euphorbias, whose juice temporarily blinds you: in twining lianas that trip you and in 'wait-a-bit' thorns that strip you. We cut, hacked and slashed our way through this nightmare vegetation. When lucky, we found a hippo path which obviated crawling on all fours.

Emson, who admitted that he had never been in a local canoe on a local island because neither harboured cows, announced after two bottles of beer and lunch that he would accompany me on my afternoon jaunt to Mugasiro Island. All went well until we were half-way across when he developed hiccups. When he 'hicked' water splashed in on one side, and on the other when he 'cupped'. We swore at him. We baled. He got worse so we baled harder. Within two hundred yards of the island he sank us. We swam ashore, not lingering. The shock cured him. Feeling wet and miserable, I left him sleeping in the sun and wandered off through the forest. Half-way up the slope, I had just found a good tsetse breeding site, when I heard a large animal crashing down through the thicket. I leapt behind a tree as an enormous crocodile whizzed past, cutting a swathe through the undergrowth. Presumably she had been laying her eggs.

140

Musoma was a lovely spot, with a 'sea' breeze: the boat called every 10 days and provided easy access to Uganda: in the hinterland there were the Serengeti plains, swarming with game, and it was so atypical of the Tanganyika I knew. I decided that it would be a good place to be transferred to, should the occasion arise.

CHAPTER 13

Our Last Months in Tanganyika Territory

On April 26th I entrained at Mwanza, reached Dodoma at midnight, and met the train from Dar es Salaam at 8 a.m. to be reunited with Wendy and meet my son and Katherine. It was good to have Musa back again.

We decided to set off for Kondoa, as the Dodoma 'hotel' was verminous. "Fat Fatima" was overladen with five adults and their loads – and the road was dreadful. For two hours we were bogged down, axle deep. There were only two boys, so we could not lift the car. We had to jack her up, insert short planks under the back wheels, repeat the performance for the front wheels, push the car the length of the planks, some four feet, and repeat the operation time after time until we got onto harder ground. We could not have managed without Katherine to take turns in holding the baby as we bumped along, and to help feed him – Musa, Arser and I were plastered with mud. We reached Kondoa at dusk – 10 hours to cover 100 miles. We spent the next day resting.

We set off for Kikori, another awful journey: we sank axle deep four times. What with tsetse trying to eat the baby, heat, rain and mud it was an unpleasant trip. It took us 6½ hours to cover 50 miles; we averaged 8 miles to the gallon of petrol. Katherine was the only person who had enjoyed the journey. She had been to India, Peru, Rhodesia and Kenya and loved bush life; she was to be much impressed by our nightly lion chorus.

For a long time Victor had had a lion cub called 'Simba'; it started off as being a delightful pet but was becoming embarrassingly large – at least for me. On Saturday afternoons Victor and I would go and shoot an animal for the camp. Simba would trot along behind Victor and periodically stand on his hind-legs and put his forepaws on his shoulders. Victor, a very burly man, would say, "down Simba, down" shrugging his shoulders, and the cub would drop to the

142

ground. But when he tried it on me I would collapse under the weight; I did not really like it. Then on one of our Saturdays, Victor shot a zebra through the shoulder; it fell and struggled to get up, but couldn't. Simba streaked towards it and then started playing with it like a cat with a mouse. Victor called him off, but he would not come. Victor tried to catch him by his collar, and he turned on Victor growling horribly. I shot the zebra through the head. We then had to sit and wait for half an hour until Simba became bored and joined us. I told Victor that he must get rid of Simba, as the askaris had started complaining. He agreed.

There then followed a period when Victor received letters from India in embossed envelopes from the Maharajas of X, Y and Z – all declining Captain Findlay's kind offer of a lion cub. Eventually, he got an acceptance from a zoo in South Africa – Port Elizabeth, I think. But how to get the cub there? That was the problem. Victor, knowing that Cecil Stiebel was shortly going on leave and was first visiting his mother in South Africa, went into Kondoa for a weekend, waited at a drinks party until everyone was feeling happy and there was a pause in the conversation, and then said 'Extraordinary how gutless people have become since the war. I can't find anyone willing to take my little lion to Port Elizabeth zoo; they want him badly.' Needless to say, Cecil, who never refused a challenge, agreed providing it was only one lion. The subsequent saga went something like this. –

Kondoa	Porters loading lorry drop crate when lion roars.
Train	Guard refuses lion unless Cecil sleeps in van. He does.
Dar	He visits Secretariat with lion on lead.
Voyage	Lion exercised daily on deck by Cecil armed with whip: passengers gawp. Crew member alerts Port Elizabeth press by radio.
Port Elizabeth	Local Press. 'Captain Stiebel, famous lion tamer, presents zoo with lion.'

The following is an extract from a letter which Wendy wrote to my mother on June 12th 1932.

'Last night we had a very tiring time. At 9.30 p.m. we went to bed as usual and at midnight Musa woke us up with a letter from Bill Potts to say that he was stuck, with a broken car, near Kisesse, 20 miles away.

It meant Tam turning out and fetching him, while I held the fort with a loaded shotgun (though I should have been far too frightened to use it). The lions were roaring all round and Tam saw

143

one on the road. They got back at 3.30 a.m. but we got no more sleep as it was nearly dawn and the cocks began to crow. We are very tired today.'

The game had started to come in from the Masai steppe, in search of water, and the lions were very noisy. They had begun to get on Katherine's nerves: she would put her hand to her heart and say "What's that?" and I would reply "Nothing – only a harmless lion." One morning, she appeared, looking poorly, and said that she had had a dreadful night – a lion had been breathing heavily under her window. I said "Nonsense! Katherine, you must pull yourself together. Come and see for yourself." We went out. There was a flower bed under her window. It was full of pug marks, and the ground flattened where a lion had lain down.

There was no question of shooting the lions, as I had been made an Honorary Game Ranger after I had got my research area prescribed as a reserve. Further, I had learnt to like lions, but hate leopards. My views would have been very different had we had a man-eater about, but we had never had a case of anyone being mauled and our lions were a feature of Kikori – rather like having peacocks on the demesne, noisy but beautiful.

On Sundays we had a regular routine. Musa, who was very proud of the baby having been at the scene of action in Dar, would lead the procession, holding the baby in a cardboard box, and saying that he had grown much heavier. We would go down the rocky hill to the laboratory where the baby would be weighed and the figure entered on the graph.

We joined a small party of friends for the Bank Holiday weekend at Bereku – a small village on top of our escarpment on the Cape to Cairo road. The Wa-ufiome were holding a great feast and dance. This tribe occupies a large territory and here, living in lovely open woodland, they were a very different people from the spirit-ridden folk on the north-eastern side of the mountain. They lived in the 'tembe' type of huts with flat, mud roofs, such as my Kandaga hut – 'Zebra Hall', p. 33.

The plump-faced maidens were dressed in all their finery for the feast, and wore only skirts. Along their fore-arms and just above the elbows were close-fitting coils of thick, highly-polished brass wire – like springs: similar coils encircled the neck, but protruded to form a ruff of concentric circles which projected 3–4 inches from the skin: long blue necklaces hung down between the breasts and a thin white band encircled the forehead. They were most decorative. On each

144

side of the spine some had down-slanted rows of cicatrices – like the barbs of an arrow. The older women wore shawls around their shoulders.

The men were lithe and agile. They wore a sarong, tucked in many times round the waist, they had bead necklaces hanging down their backs; all carried spears – some 6 feet long – and circular shields with a central boss. We watched many a single combat, fought by the experts with a naked spear in the right hand and in the left a stick and shield with which to beat down or deflect the opponent's spear. The fights took place at tremendous speed, first crouched, and near-kneeling, a few feet apart and then nearly foot to foot with legs wide apart and each thrusting savagely. Their speed of eye must have been fantastic for no accident to occur.

Soon after we got back, Swynnerton arrived unexpectedly with Harold Lloyd who was to take over from me. Swynnerton wanted me to make an eight mile long fire-break at once, which was inopportune as I should have been handing over to Lloyd before going on leave; he also had a luminous idea that next tour I should study a certain waterhole in a remote area, only approachable by a dry season road – not an ideal setting for wife and baby.

On my last morning at Kikori I was writing in my laboratory when I heard shrieks from some women on the road; I looked up – a lion was chasing a lioness. I then heard a yell from the clerk, who was on the point of walking across from his office when the lions shot past him. They finally mated some 20 yards from my laboratory window. How typical of Kikori – that zoo without bars.

We were due to sail on the Llanstephan Castle on September 11th 1932, and drove down to Dar es Salaam with Katherine, Musa and Arser, as I wanted to get the car over-hauled. Little did we know when they saw us off that we would never meet again.

CHAPTER 14

A Change of Course

When I had been interviewed in the Colonial Office for the post in Tanganyika, I was told that it was expected that the appointment would be made pensionable in about 6 months time; it still wasn't after over five years, and meanwhile we had got married and had started a family. As soon as we landed, I went round interviewing the few influential friends I had in search of a pensionable job, but vacancies were not being filled owing to the depression.

We had one unforgettable week. Cecil Stiebel was on leave, he had become engaged and wanted a chaperone. 'Would we join them in a small hotel at Westcliffe-on-Sea?' Of course we agreed to do so. It was an excellent area in mid-winter, as it was empty. We roller-skated on a deserted rink in Southend, we went beagling in the adjacent countryside, and went to the cinema at least once a day. Everywhere we went, Cecil was well-known.

One night, we went upstairs together on our way to bed to find the landing full of smoke. We could see no fire but notified the proprietress who joined us in sniffing round like a pack of hounds; we spotted the source – smoke was coming from under a bedroom door. She knocked on the door, the irate occupier opened it:

'What do you want?'

'Your room is on fire.'

'It isn't. I am burning old socks in the grate.'

Silly incidents like this always happened with Cecil.

One evening he drove us up to London to visit his old friends in the night-club world whom he had not yet had time to see. Everywhere he took us, the girls would throw their arms round his neck and say 'Felix, darling, where *have* you been?' He was always known as 'Felix', after the cat of that name in the current cinema cartoon, because like that cat he was always walking and never sat down.

Cecil was a tremendous character, and the natives loved him – 'the great white bull'. Once I walked into his office in Kondoa, when his boss, the District Officer, was away on tour. There was Cecil, sitting in the official chair, strumming on his mandolin to a delighted party of local chiefs. He died many years ago, but we still keep in touch with Jane, his wife.

After some months, the Colonial office offered me the permanent and pensionable post of Medical Entomologist, Nigeria, with secondment to the Sleeping Sickness Service. I accepted because of the security offered and the attractiveness of the job, even though it entailed a big drop in salary and separation from our child. (In those days children of government servants were quite rightly banned. The reputation of the 'white-man's grave' still lingered. It was not until after the Second World War that the ban was lifted as a result of the discovery of newer drugs – more especially of the anti-malarials.) One big advantage was that a tour of service was only 18 months, instead of the 30 months in East Africa; the voyage was also much shorter – a fortnight, instead of 5 weeks.

The job was most attractive in that I would virtually get a free hand as to the direction of my research, and would be expected to visit the areas where the doctors had found the highest incidence of the disease in order to locate the sources of infection which must be eradicated. At last, I would be able to associate my research with the practical problem, and co-operate with the doctors who knew little about tsetse, whilst I knew nothing about the disease – a perfect basis for co-operation. Further, Dr. W. B. Johnson, who had spent a week at Kikori and who I so greatly admired, was now Director, Medical Services, Nigeria.

I was soon to lose touch with my friends in Tanganyika because in Kipling's words ' . . . East is East and West is West and never the twain shall meet'; that was the situation in Africa until after the war, when air-travel enabled international scientific conferences to be held and one met one's fellow-workers.

I was to spend the next 26 happy years in Nigeria.

147

Addendum

On June 8th 1938, Swynnerton, Burtt, and the pilot of a Leopard Moth were killed in an air crash, half way between Singida and Dodoma. It was three days before the wreckage was found by R.A.F. planes from Kenya. Swynnerton was on his way down to Dar es Salaam to receive his C.M.G. Noel tells me that it was believed that the pilot was blinded by the rising sun as he flew very low to enable Burtt to photograph a rhino, as was inferred from a film in a camera found on the site. Their deaths were a very great loss to science.

In about 1896, when 19 years old, Swynnerton went out to Rhodesia and became supervisor of a large farm in the Melsetter district near the Portuguese border. He was a dedicated naturalist: of some 1,100 plant specimens which he sent home for identification, 190 were new to science. He also investigated the theories on animal and insect coloration in relation to the unpalatability of certain insects as food for birds. In 1918 he started studying tsetse flies in Portuguese East Africa and in the following year became the first Game Warden of Tanganyika Territory. He combined the task of game preservation with studies on the relationship between game and tsetse. In 1921, he concluded that two thirds of the Territory was infested by these insects. In 1929 he left the Game Department to become the Director of the new Department of Tsetse Reclamation and Research which he had created. He combined his great knowledge of natural history with an infectious enthusiasm which he communicted in full measure to his staff, and to those who controlled the giving of grants in aid of research and reclamation. He was a truly great and lovable man, but I *did* find him exhausting.

Burtt, joined Swynnerton in 1925 and was one of the first members of Swynnerton's research team. Burtt, the son and cousin of well-known botanists, was reputed to have a more intimate knowledge of

148

the vegetation of Tanganyika Territory and probably of tropical East Africa, than any other living botanist – even though he was only 36 years old when he was killed. Already one genus and eleven species of plants new to science had been named after him. But it was his unfailing kindness, charm, and eccentricity which made him loved by all. In early 1938 he spent a week with us in bush in Northern Nigeria, together with Jock King who kindly lent me the obituary notices upon which some of the above is based. On leaving us they drove across Africa to Tanganyika Territory – no mean feat in those days.

Both Swynnerton and Burtt are buried on a rocky hill at Old Shinyanga, the then headquarters of the tsetse research work. It is good to think of them both overlooking the Africa they loved so dearly.

A few years later Victor Findlay, seriously injured by a rhino, was carried for some days in a hammock to the nearest hospital, where he died.*

Noel Vicars-Harris, having been Assistant Director of the Department of Tsetse Research, left in 1937 to become Secretary of the Lands and Mines Department.

Noel and I are probably the last survivors of Swynnerton's original team.

*He was also buried beside Swynnerton and Burtt.

149

Glossary

A.D.O.	— Assistant District Officer	*mbuga*	— a low-lying area where water collects during the rainy season
askari	— a soldier in, or retired from, the King's African Rifles	*mbwa*	— a dog
		mbweha	— a jackal
banda	— a thatched hut	*miombo*	— a common type of woodland, *Berlinia* and/or *Brachystegia*
boma	— a district or divisional head-quarters, literally a thorn fence as used in the early days		
		mtoto	— a child
		mwanangwa	— a village headman
		mzee	— an old man
boy	— a servant, irrespective of age, e.g. house-boy, garden-boy	*nagana*	— animal trypanoso-miasis, the disease transmitted by tsetse flies to domestic animals in Africa
bwana	— the master		
D.O.	— District Officer, The administrator of a district	*panga*	— cutlass or machete
		P.C.	— Provincial Com-missioner, the administrator of a province
duka	— a shop or store		
faru	— a rhinoceros		
H.E.	— His Excellency, i.e. The Governor of the Territory	*safari*	— expedition or journey
		shamba	— a farm or culti-vated land
jumbe	— a chief		
K.A.R.	— King's African Rifles	*shetani*	— the devil
		siafu	— biting 'driver' ants
kiboko	— a rhino hide whip	*simba*	— a lion
machan	— a shooting plat-form built up a tree	sleeping sickness	— human trypano-somiasis, the African disease

150

		transmitted by tsetse flies to man	*tembe*	—	a mud and wattle hut with a *flat* mud roof
sundowner	—	evening drinks starting at sunset or a drinks party	tour	—	period of service in Africa between home-leaves
Swahili	—	universal East African language, a lingua franca for the many tribal languages	*Wa-ufiome*	—	members of the Ufiome tribe (similarly Wasandawe, Wambulu, etc.)

Bibliography

BACH, Eve. (1934). *The Youngest Lion*. London : Hutchinson.
BAILEY WILLIS (1930). *Living Africa*. New York : McGraw-Hill.
BLAYNEY PERCIVAL, A. (1930). *A Game Ranger's Note Book*. London : Nisbet.
BRODHURST-HILL, E. (1936). *So This is Kenya*. London : Blackie.
CARNOCHAN, F. G. and ADAMSON, H. C. (1935). *The Empire of the Snakes*. London : Hutchinson.
DENEYS REITZ (1933). *Trekking On*. London : Faber and Faber Ltd.
HUXLEY, Julian. (1931). *Africa View*. London : Chatto and Windus.
IONIDES, G. J. P. (1964). *A Hunter's Story*. London : W. H. Allen.
LAWICK-GOODALL, J. Van and H. Van (1970). *Innocent Killers*. London : Collins.
LEAKEY, L. S. B. (1937). *White African*. London : Hodder and Stoughton.
MASEFIELD, John (1927). *Multitude and Solitude*. London : Jonathan Cape
NASH, T. A. M. (1969). *Africa's Bane : The Tsetse Fly*. London : Collins.

INDEX

153

154

155